PATTERNS AND FUNCTIONS

GLOBE FEARON EDUCATIONAL PUBLISHER
A Division of Simon & Schuster
Upper Saddle River, New Jersey

Executive Editor: Barbara Levadi
Editors: Bernice Golden, Lynn Kloss, Bob McIlwaine, Kirsten Richert, Tom Repensek
Production Manager: Penny Gibson
Production Editor: Walt Niedner
Interior Design: The Wheetley Company
Electronic Page Production: The Wheetley Company
Cover Design: Pat Smythe

Reviewers:
Lorraine Carlozzi
Assistant Principal for Mathematics
Midwood High School, Brooklyn, NY

Judy Walsh, B.S., M.S.
Assistant Principal for Mathematics
New Dorp High School, Staten Island, NY

Printed in the United States of America 1 2 3 4 5 6 7 8 9 10 99 98 97 96 95

ISBN 0-8359-1542-5

CONTENTS

TO THE STUDENT

Access to Math is a series of 15 books designed to help you learn new skills and practice these skills in mathematics. You'll learn the steps necessary to solve a range of mathematical problems.

LESSONS HAVE THE FOLLOWING FEATURES:

❖ Lessons are easy to use. Many begin with a sample problem from a real-life experience. After the sample problem is introduced, you are taught step-by-step how to find the answer. Examples show you how to use your skills.

❖ The ***Guided Practice*** section demonstrates how to solve a problem similar to the sample problem. Answers are given in the first part of the problem to help you find the final answer.

❖ The ***Exercises*** section gives you the opportunity to practice the skill presented in the lesson.

❖ The ***Application*** section applies the math skill in a practical or real-life situation. You will learn how to put your knowledge into action by using manipulatives and calculators, and by working problems through with a partner or a group.

Each book ends with ***Cumulative Reviews***. These reviews will help you determine if you have learned the skills in the previous lessons. The ***Selected Answers*** section at the end of each book lists answers to the odd-numbered exercises. Use the answers to check your work.

Working carefully through the exercises in this book will help you understand and appreciate math in your daily life. You'll also gain more confidence in your math skills.

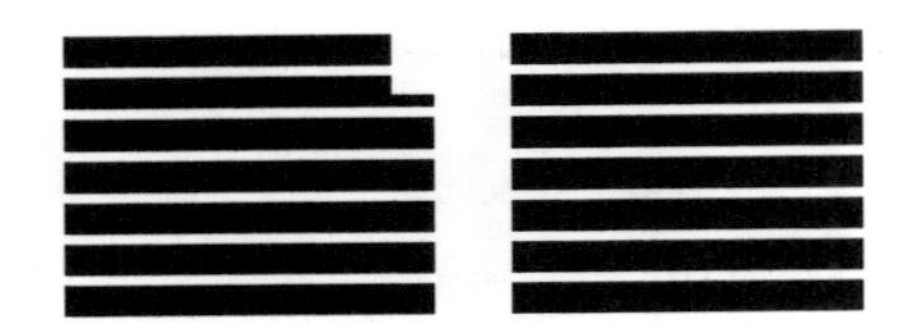

FINDING A NUMBER PATTERN

Vocabulary

pattern: items arranged or repeated in an ordered way

number pattern: a group of numbers in an order that follows a rule

When we find and use **patterns**, we can discover efficient ways for calculating. Finding a rule for a **number pattern** is like solving a puzzle.

Tanika and Paul restock the shelves of their bakery according to a schedule. The schedule begins at 6 A.M. with the first batch of fresh bread. They restock their shelves with fresh bread at 8 A.M., 10 A.M., 12 P.M., and so on. How many more times will Tanika and Paul restock their shelves with fresh bread before the bakery closes at 5:30 P.M.?

To figure out the pattern, list the numbers in the schedule.

6 A.M. → 8 A.M. → 10 A.M. → 12 P.M. → ?
(+2, +2, +2, +2)

Reminder

When adding hours, counting starts over at 12 noon and 12 midnight.

Notice that the amount of time that passes between each batch is always the same.

This pattern uses addition. The rule is add 2. Tanika and Paul restock their shelves with bread at 2 P.M. and 4 P.M. They will restock their shelves two more times before the bakery closes at 5:30 P.M.

Guided Practice

1. The in-store bakery employees at a local supermarket put batches of fresh bread on the shelves at 7 A.M., 10 A.M., 1 P.M., 4 P.M., and so on.

a. How much time passes between batches? 3 hours

b. What is the rule for this pattern? ______

c. How many more batches of fresh bread will the bakery employees place on the shelves before the store closes at 11 P.M.? ______

Exercises

Find the rule for each number pattern. Write the next three numbers in the pattern.

2. 3, 5, 7, 9 , ____, ____, ____

3. 14, 17, 20, 23, ____, ____, ____

4. 1, 11, 21, 31, ____, ____, ____

5. 4, 9, 14, 19, ____, ____, ____

6. 60, 64, 68, 72, ____, ____, ____

7. 2.5, 3.4, 4.3, 5.2, ____, ____, ____

8. $4\frac{1}{2}$, 6, $7\frac{1}{2}$, 8, ____, ____, ____

9. 12, 25, 38, 51, ____, ____, ____

Application

10. At an Italian bakery, a pizzel costs $.50. It costs $1 for two pizzels, $1.50 for three pizzels, and $2 for four pizzels. How much will 13 pizzels cost? Discuss with a partner an efficient way of finding the cost for any number of pizzels and write your strategy.

____

____

____

11. Create your own number pattern. Exchange patterns with a partner and write the rule for your partner's pattern.

____

____

____

MULTIPLES

Vocabulary

multiple: the product of a whole number and any other number

Janeen is saving to buy a mountain bike. She saves $9 each week. How much money will she save in seven weeks?

We can count by nines 7 times.

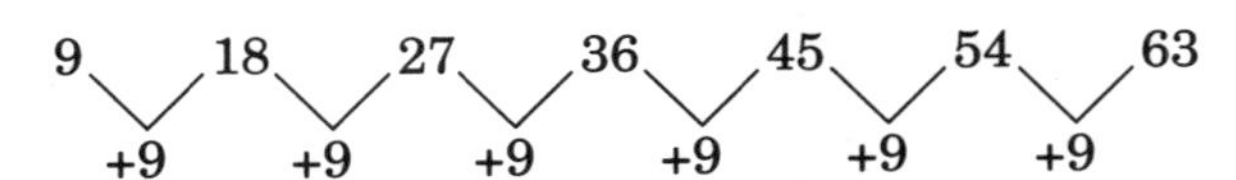

The numbers 9, 18, 27, 36, 45, 54, and 63 are the first seven **multiples** of 9.

We can also set up a table showing each week she saved $9.

Weeks	1	2	3	4	5	6	7
Money Saved	$9	$18	$27	$36	$45	$54	$63

By looking at the table, we can see that the seventh multiple of 9 is 63 (7×9).

Janeen will save $63 in seven weeks.

Guided Practice

1. Look at the numbers below.

6, 12, 18, 24

a. These numbers are the first four multiples of _6_.

b. What are the next three multiples?

6, 12, 18, 24, _30_, ____, ____

c. What is the 13th multiple of 6? ________

Exercises

List the next three multiples. Then find the 20th multiple.

2. 5, 10, 15, ____, ____, ____

20th multiple: __________

3. 13, 26, 39, ____, ____, ____

20th multiple: __________

4. 8, 16, 24, ____, ____, ____

20th multiple: __________

5. 10, 20, 30, ____, ____, ____

20th multiple: __________

6. 4, 8, 12, ____, ____, ____

20th multiple: __________

7. 11, 22, 33, ____, ____, ____

20th multiple: __________

Application

8. Jevon's Bike Rentals charges $12 per hour to rent a mountain bike.

a. Complete the table.

Hours	1	2	3	4	5
Cost	$12	$24			

b. How many hours could you rent a bike if you have $168?

____

9. Describe how you found your answers to Application 8.

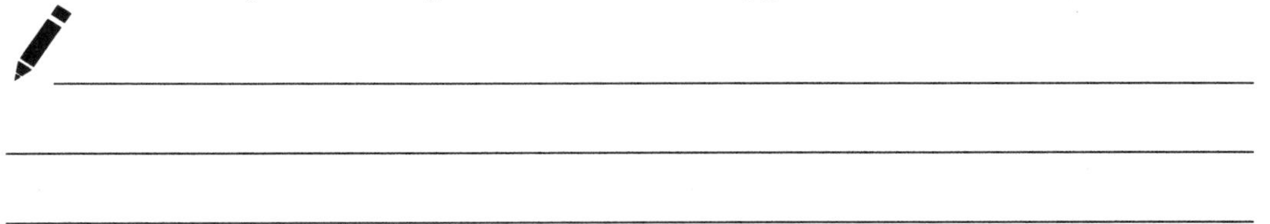

10. What number is a multiple of every number? __________

CONSTRUCTING PATTERNS BASED ON A RULE

Pierre wants to rent a snowmobile. The rate is $8 per hour and a $30 deposit. How much will it cost Pierre to rent a snowmobile for 1 hour?

Pierre uses the following rule to find out the cost.

$8 times the number of hours plus $30 deposit = Cost

↓ ↓ ↓ ↓ ↓ ↓ ↓

$8 \quad \times \quad h \quad + \quad 30 \quad = \quad C$

You can create a pattern using the rule. Write the number of hours in order. Substitute the number of hours for h.

$$8h + 30$$

$$8 \times 1 + 30$$

$$38$$

Reminder

$8h$ means $8 \times h$

Hours (h)	1	2	3	4	5	6
Cost ($8h + 30$)	$38					

It will cost Pierre $38 to rent a snowmobile for one hour.

Guided Practice

1. How much will it cost Pierre to rent a snowmobile for 2 hours?

a. Write the rule. $8h + 30$

b. Substitute 2 for h. $8 \times 2 + 30$

c. Solve to find the cost. Cost = ____________

2. Finish completing the table above.

a. Substitute 3 for h. Cost = ____________

b. Substitute 4 for h. Cost = ____________

c. Substitute 5 for h. Cost = ____________

d. Substitute 6 for h. Cost = ____________

Exercises

Complete the table using the rule. The first one is started for you.

3. Rule: $3n + 2$

n	1	2	3	4
$3n + 2$	5			

4. Rule: $5m - 2$

	1	2	3	4

5. Rule: $9a + 7$

	1	2	3	4

6. Rule: $4b - 4$

	1	2	3	4

7. Rule: $6r - \frac{1}{2}$

	1	2	3	4

8. Rule: $0.5p + 2$

	1	2	3	4

Application

9. Monique needs to have some repairs done on her snowmobile. The service station charges $27 per hour. The mechanic took 3 hours and 30 minutes to fix the snowmobile. Parts came to $53.50. How much will it cost Monique to repair her snowmobile? Use the table below to help you solve the problem.

a.

Hours (h)						
Cost ($27h + 53.50$)						

b. Monique's cost is ____________.

10. Describe how you found the answers to Application 9.

____

____

____

WRITING THE RULE FOR A NUMBER PATTERN

Sometimes it is more efficient to write and use a rule for a pattern than it is to continue the pattern to solve a problem. The first five terms in a pattern are shown below. What is the 90th term?

1, 4, 7, 10, 13, . . .

You can write a rule for the pattern to find the 90th term.

The first characteristic of the pattern is that each number increases by 3.

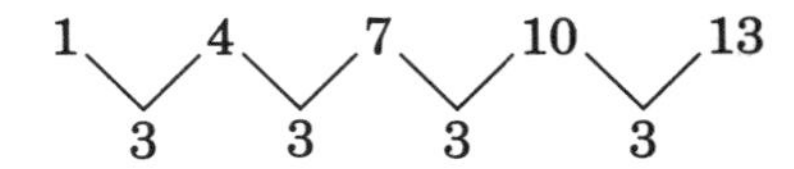

Since the pattern is increasing by 3, the multiplier is 3. Start a rule with $3n$.

Term Number (n)	1	2	3	4	5
Rule: $3n$	1	4	7	10	13

Next, substitute 1 for the first term number.

$$3n = 3 \times 1 = 3$$

Since the first number in the pattern is 1, this is not the rule. If you subtract 2 from 3, you get 1. So the rule could be $3n - 2$. Test this rule with the second term.

$$3n - 2 = 3 \times 2 - 2 = 4$$

Test the third term. $3 \times 3 - 2 = 7$

Test the fifth term. $3 \times 5 - 2 = 13$

The rule works. Use the rule to find the 90th term. Let $n = 90$.

$$3 \times 90 - 2 = 268$$

The 90th term is 268.

Guided Practice

1. Use the number pattern: 5, 9, 13, 17, . . .

   a. By how much are the numbers increasing? 4

   b. What is the multiplier? 4

   c. Let n stand for the term number. Write the rule so far. $4n$

   d. Substitute 1 for n. What do you do to 4×1 to get 5? ______

   e. Write the rule for the pattern. ______

   f. Try the rule on the second term. ______

   g. What is the 70th term? ______

Exercises

The first four terms of each pattern are given. Write the rule for the pattern. Then find the 80th term.

2. 5, 7, 9, 11, . . .

   Rule: ______

   80th term: ______

3. 11, 21, 31, 41, . . .

   Rule: ______

   80th term: ______

4. 3, 7, 11, 15, . . .

   Rule: ______

   80th term: ______

5. 6, 12, 18, 24, . . .

   Rule: ______

   80th term: ______

6. 2, 9, 16, 23, . . .

   Rule: ______

   80th term: ______

7. 4.5, 7.5, 10.5, 13.5, . . .

   Rule: ______

   80th term: ______

Application

8. With a partner, play "Guess My Rule." Think of a rule like the ones you wrote in this lesson. Your partner gives you a number to substitute into your rule. Tell your partner the result. Your partner can then try to guess the rule or give you another number. After your partner guesses the rule, switch roles. Describe any strategies you or your partner used to guess the rule.

______

______

______

GEOMETRIC PATTERNS

Vocabulary

diagonal: a line segment that joins two vertices of a polygon but is not a side

Reminder

A polygon having all sides equal in length and all angles equal in measure is a *regular polygon*.

Discovering a pattern can be exciting. There are many patterns in regular polygons. Jamika was exploring **diagonals**. She thought there might be a relationship between the number of sides in a regular polygon and the number of diagonals.

To find out, she made a table.

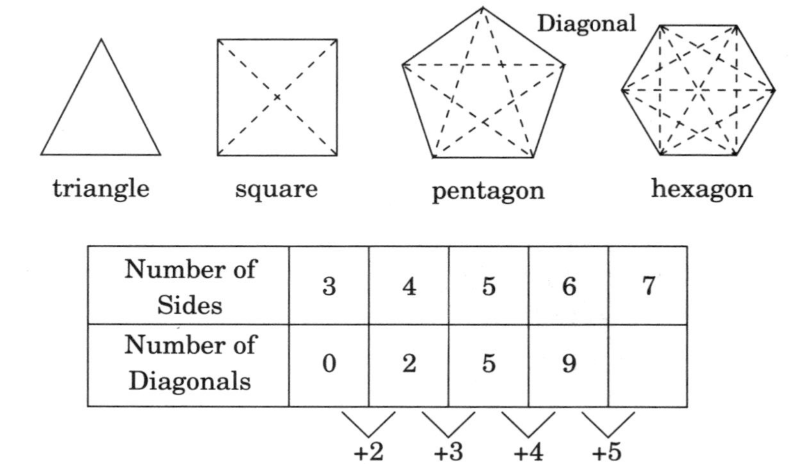

Number of Sides	3	4	5	6	7
Number of Diagonals	0	2	5	9	

Jamika noticed the pattern shown below the table. Jamika continued the pattern by adding 5, and found that a 7-sided regular polygon has 14 diagonals.

Guided Practice

1. How many diagonals are in an 8-sided regular polygon?

a. How many diagonals are in a 7-sided regular polygon? ____14____

b. What number will you add to continue the pattern? ________

c. An 8-sided polygon has ________ diagonals.

2. Look for a pattern.

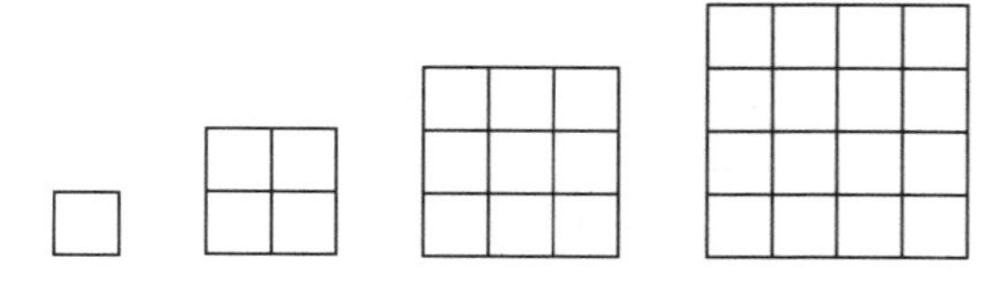

a. How many little squares are in each figure?

____

b. Draw the fifth figure. How many little squares are in the figure? ____________

c. Without drawing the sixth figure, how many little squares are in the sixth figure? ____________

d. What is the pattern? ____________

Exercises

Look for a pattern. Draw the next shape. Describe the pattern.

3.

4.

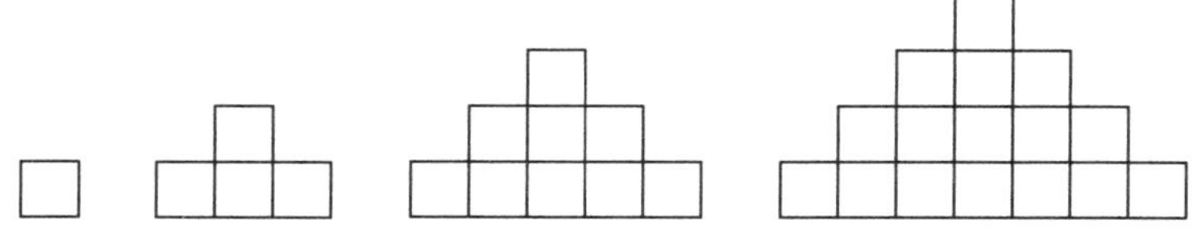

5.

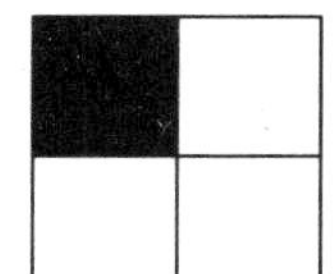
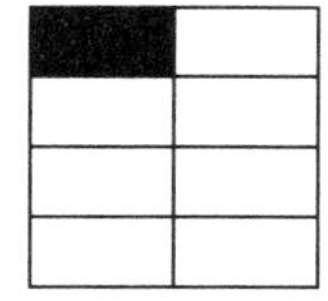

Application

Work with a partner to solve this problem.

6. Find the number of diagonals that can be drawn from any one vertex of any 8-sided polygon.

a. Draw several polygons with different numbers of sides. See how many diagonals you can draw from any one vertex.

b. Make a table from your results. Look for a pattern to solve the problem.

Number of Sides	3	4	5	6	7	8
Number of Diagonals						

____

____

PRIME NUMBERS

Vocabulary

prime number: a number with only two factors, itself and 1

composite number: a number that has more than two factors

Reminder

Factors are numbers that are multiplied together to form a product.

Bruno is arranging tables to form a rectangle. He notices that the only way he can arrange 13 tables is in 1 row of 13 tables.

The number 13 is a **prime number**. A prime number has only two factors, 1 and itself. Numbers with more than two factors are called **composite numbers**.

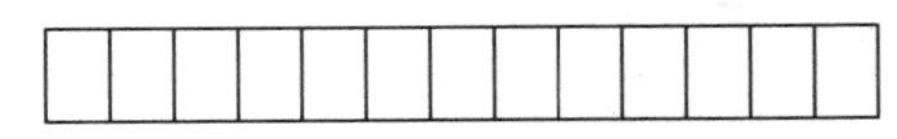

13 x 1

The table below shows all the factors of the numbers 2 to 14. The numbers with only two factors are prime numbers. The numbers with more than two factors are composite numbers.

										12		
										6		
				6		8		10		4		14
		4		3		4	9	5		3		7
2	3	2	5	2	7	2	3	2	11	2	13	2
1	1	1	1	1	1	1	1	1	1	1	1	1
2	**3**	**4**	**5**	**6**	**7**	**8**	**9**	**10**	**11**	**12**	**13**	**14**

Guided Practice

1. A prime number has how many factors? 2

2. List all the prime numbers shown in the table above. ______

3. ______ numbers are numbers that have more than two factors.

4. List all the composite numbers shown in the table.

______

5. Circle the prime number at the right. 15 21 27 31

Exercises

6. Find all the prime numbers less than 100.

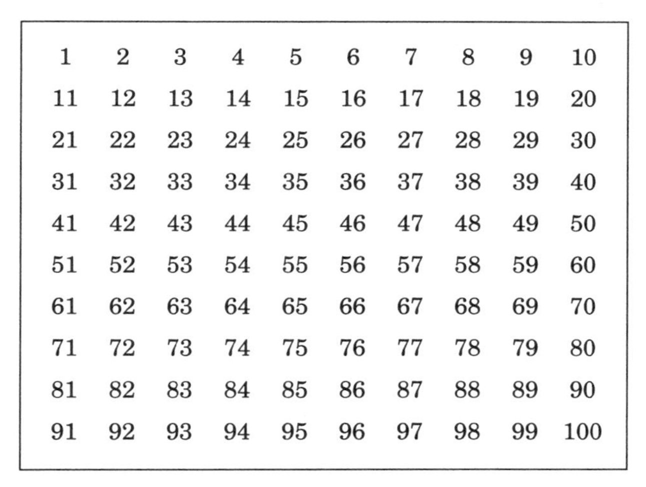

1	2	3	4	5	6	7	8	9	10
11	12	13	14	15	16	17	18	19	20
21	22	23	24	25	26	27	28	29	30
31	32	33	34	35	36	37	38	39	40
41	42	43	44	45	46	47	48	49	50
51	52	53	54	55	56	57	58	59	60
61	62	63	64	65	66	67	68	69	70
71	72	73	74	75	76	77	78	79	80
81	82	83	84	85	86	87	88	89	90
91	92	93	94	95	96	97	98	99	100

a. Draw a circle around the number 2.

Cross out all the other multiples of 2.

b. Draw a circle around the number 3.

Cross out all the other multiples of 3.

c. Do the same for 5 and 7.

d. Draw a circle around the remaining numbers that are not crossed out.

e. The circled numbers are prime numbers.

Use the numbers above. Circle *true* or *false*.

7. The number 77 is prime.	true	false
8. The number 87 is composite.	true	false
9. All odd numbers are prime.	true	false
10. The only even prime number is 2.	true	false

Application

11. Work with a partner to make rectangles using squares. Cut 24 (1 in. × 1 in.) squares from paper. Start with two squares. How many rectangles can you make? You can only make one rectangle (1 × 2). Do the same using three squares, four squares, and so on until you use 24 squares. Make a table or a list of your results.

12. What do you notice about the number of rectangles that you can make?

____

PRIME FACTORS

Vocabulary

prime factorization: the expression of a number as the product of prime factors

Reminder

A prime number has only two factors, itself and 1. A composite number is any number that has more than two factors.

Rajeev and Susan were asked to write the number 60 as the product of factors. Factors are usually whole numbers that are multiplied.

Rajeev	Susan
$60 = 6 \cdot 10$	$60 = 2 \cdot 2 \cdot 3 \cdot 5$

Both answers are correct. The difference between Rajeev's factors and Susan's factors is that Susan used prime factors.

You can write any composite number as the product of prime factors. Susan found her factors by drawing a diagram called a factor tree.

60

(3) × 20

4 × (5)

(2) × (2)

$60 = 2 \times 2 \times 3 \times 5$

Use the factor tree to find the **prime factorization**, or prime factors, of 60.

Step 1 Factor the original number into two factors.

Step 2 If one of the numbers is prime, circle it.

Step 3 If a number is not prime, factor it again.

Step 4 Write the circled numbers as a product of prime factors, from least to greatest.

Guided Practice

1. Write the prime factorization for 40.

40

(5) × A

(2) × B

(C) × (2)

a. $5 \times A = 40$, $A =$ ___8___

b. $2 \times B = A$, $B =$ ________

c. $C \times 2 = B$, $C =$ ________

d. What is the prime factorization of 40?

________ × ________ × ________ × ________

Exercises

Use factor trees to find the prime factorization.

2. 18

3. 30

4. 54

5. 150

Application

6. Find the prime factorizations for 16, 24, 40, 56, 72, and 88. Finish the table below. The first row is done for you. Tell about any patterns you find.

Number	Prime Factorization				
16	2	2	2	2	
24					
40					
56					
72					
88					

7. Describe any patterns you notice in the prime factors of 16, 24, 40, 56, 72, and 88.

___

___

___

FINDING NUMBER PATTERNS BASED ON MULTIPLICATION AND DIVISION

A youth group trip was canceled because of rain. As president of the group, Jenny formed a "phone tree" to tell members quickly. At each stage of the phone tree, a person calls two others. At Stage 1, Jenny calls Betty and Tom. At Stage 2, Betty calls Amil and Roseanna, and Tom calls Mark and Colette. If the pattern continues, how many people will be called at Stage 5?

This problem is easier to understand if the information is organized in a tree diagram.

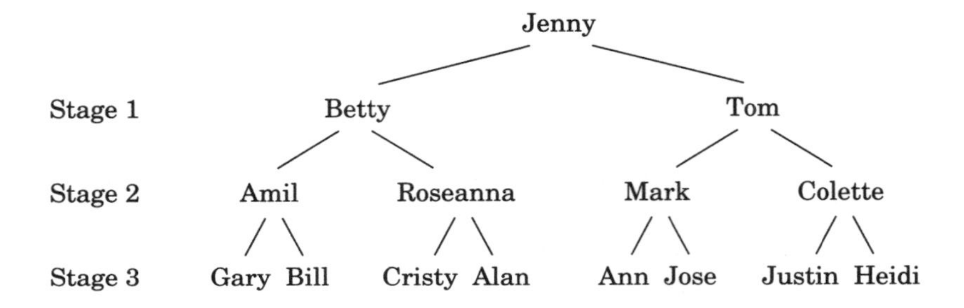

You can also use a table to organize the data.

Stage Number	1	2	3	4	5
Number of People Called	2	4	8	16	32

$\times 2$ $\quad \times 2$ $\quad \times 2$ $\quad \times 2$

The number of people called at each stage is multiplied by 2. By Stage 5, 32 people will have been called.

Guided Practice

1. If the phone tree pattern continues, how many people will be called by Stage 7?

a. How many were called by Stage 5? ___32___

b. How many at Stage 6? ___32___ $\times$ 2 = ________

c. How many at Stage 7?

_______ $\times 2 =$ _______

2. The number of people called at Stage 2 is 2×2 or 2^2. Write the number of people called at each stage as a power of 2.

a. Stage 5 _______

b. Stage 6 _______

c. Stage 7 _______

Exercises

Find the next term in each pattern. Describe the pattern. You may use a calculator to help you.

3. 3, 9, 27, 81, . . .

____

4. 400, 200, 100, 50, . . .

____

5. 5, 20, 80, 320, . . .

____

6. 7, 14, 28, 56, . . .

____

7. 810, 270, 90, 30, . . .

____

8. 0.01, 0.1, 1, 10, . . .

____

Application

9. Yianna and Oksana were given the same first three numbers in a pattern: 2, 4, and 8. However, the next two numbers in Yianna's pattern were different from Oksana's next two numbers.

Yianna's pattern
2, 4, 8, 16, 32, . . .

Oksana's pattern
2, 4, 8, 32, 256, . . .

Describe Yianna's rule and Oksana's rule. Use their rules to write the sixth term in their patterns.

____

____

____

FINDING A RULE

Vocabulary

sequence: a set of numbers arranged in a pattern

Discovering a rule for a pattern is like doing detective work. You look for clues among the numbers. Some sets of numbers do not seem to follow a pattern.

Example 1

Sumi studied the pattern below. At first, she did not see a pattern. She decided to find the difference between each term.

1, 1, 2, 3, 5, 8, 13, 21, 34, 55, . . .

0 1 1 2 3 5 8 13 21

Sumi found that after the initial zero, the number pattern for the differences was the same as the original pattern. But she still did not see the pattern. She looked closer and found a rule: each term in the pattern was found by adding the previous two terms.

$1 + 1 = 2 \quad 1 + 2 = 3 \quad 2 + 3 = 5 \quad 3 + 5 = 8$

The number pattern above is called the Fibonacci Sequence. The pattern was named after Leonardo Fibonacci, a mathematician of 13th-century Europe. A **sequence** is a number pattern with no end. The three dots, called an *ellipsis*, indicate that the numbers continue in the same pattern.

Example 2

Sumi decided to make her own pattern and call it the Sumi Sequence. She used the rule $n^2 + 3$ and made a table.

Term Number (n)	1	2	3	4	5	6	7
Rule: $n^2 + 3$	4	7	12	19	28	39	52

The Sumi Sequence is 4, 7, 12, 19, 28, 39, 52, . . .

Guided Practice

1. Use the Fibonacci Sequence.

a. What are the last two numbers shown in the pattern? 34 and 55

b. Add the numbers to find the next term in the pattern. ________

2. What is the eighth term in the Sumi Sequence?

____

Exercises

Write the next three terms in each sequence. Then state the rule.

3. 1, 4, 9, 16, 25,

________, ________, ________, . . .

Rule: ____________________

4. 10, 10, 20, 30, 50,

________, ________, ________, . . .

Rule: ____________________

5. 67, 69, 60, 62, 53,

________, ________, ________, . . .

Rule: ____________________

6. 1, 2, 2, 3, 3, 3, 4, 4,

________, ________, ________, . . .

Rule: ____________________

Use the rule to write the first six terms in a pattern. Let *n* stand for the term number.

7. $n^2 + n$

________, ________, ________, ________, ________, ________, . . .

8. 5^n

________, ________, ________, ________, ________, ________, . . .

9. The first six rows of Pascal's Triangle are shown at the right. What pattern is used to find each number in the rows? What numbers are in the 7th row?

____

1
1 1
1 2 1
1 3 3 1
1 4 6 4 1
1 5 10 10 5 1

Application

COOPERATIVE LEARNING

Work with a partner.

10. Suppose you want to make your own Fibonacci-like Sequence, but you start with the number two. The first three numbers are given. Write the next five terms in the sequence.

2, 2, 4, ________, ________, ________, ________, ________, . . .

10 DOT PATTERNS

Ivis is the manager of a supermarket. He wants to stack soup cans in layers for a display. How many soup cans does he need for the tenth layer?

To start, Ivis draws dot patterns for each layer of a four-layer display.

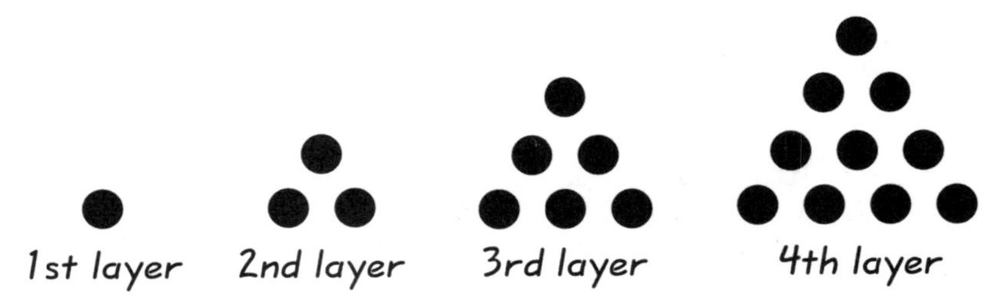

To see the pattern, Ivis makes a table.

	Number of Layers	Sum of Cans in Layer	Number of Cans in Layer
Layer 1	1	1	1
Layer 2	2	1+2	3
Layer 3	3	1+2+3	6
Layer 4	4	1+2+3+4	10

He sees that the number of soup cans in a layer is equal to the sum of the layer numbers.

So, Ivis adds the numbers from one to ten.

$$1 + 2 + 3 + 4 + 5 + 6 + 7 + 8 + 9 + 10 = 55$$

The tenth layer of the display needs 55 cans of soup.

Guided Practice

1. What is the total number of cans Ivis needs for a ten-layer display? Look for a pattern.

Number of Layers	1	2	3	4	5	6	7	8	9	10
Number of Cans in Layer	1	3	6	10	15	21	28	36	45	55
Total Number of Cans	1	4								

a. Look for a pattern. Complete the table.

b. How many soup cans does Ivis need for a ten-layer display? ____________

Exercises

Look at the dot patterns below. Draw the next dot figure for each pattern. You may want to make a table to help.

2.

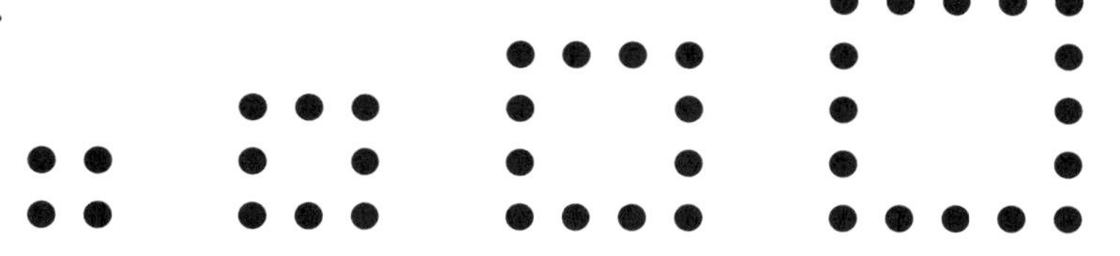

How many dots are in the tenth figure? ____________

3.

How many dots are in the tenth figure? ____________

4.

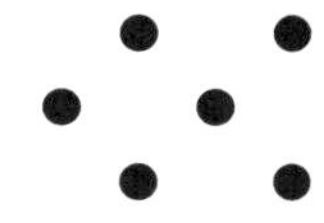

How many dots are in the tenth figure? ____________

Application

5. Ivis makes another display. He stacks boxes as shown at the right. The top layer has one box. The second layer has four boxes. The third layer has nine boxes, and so forth. Try this with cubes. How many boxes are in the tenth layer? ____________

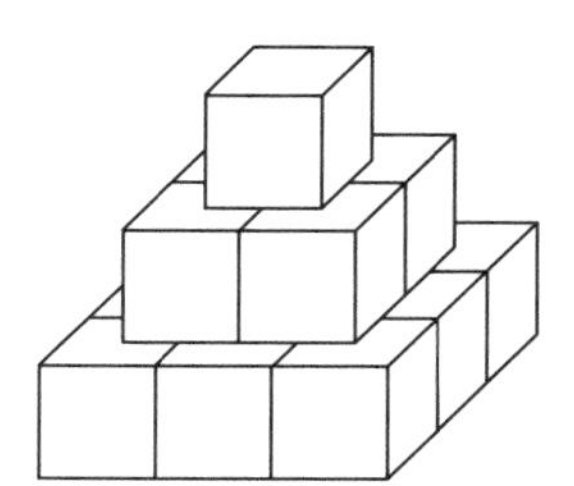

PATTERNS BASED ON HOURS OF THE DAY

Mr. Jackson takes two cold medicines at 8 A.M. If he takes medicine A every 3 hours and medicine B every 4 hours, when will he take the two medicines at the same time again?

Mr. Jackson makes two tables to show the times when he will take the cold medicines.

Medicine A	1	2	3	4	5
Time	8 A.M.	11 A.M.	2 P.M.	5 P.M.	8 P.M.

Medicine B	1	2	3	4	5
Time	8 A.M.	12 P.M.	4 P.M.	8 P.M.	12 A.M.

The pattern for medicine A can be continued by adding 3 each time. The pattern for medicine B can be continued by adding 4 each time.

The tables clearly show that Mr. Jackson will take the two medicines again at 8 P.M.

Guided Practice

1. Mr. Jackson is to take the medicines as indicated for 24 hours. When will he take the two medicines at the same time again after 8 P.M.?

a. Extend the pattern for medicine A to cover a 24-hour period.

11 P.M. - 2 A.M. - 5 A.M. - 8 A.M.

b. Extend the pattern for medicine B to cover a 24-hour period.

____

c. He will take the two medicines again at

____________.

2. Use the tables to find the number of times Mr. Jackson will take the medicines before midnight.

a. Medicine A: 6 times

b. Medicine B: ______

Exercises

3. The average person blinks about 600 times in one hour. How many times will a person blink in 3 hours? in 6 hours? in 9 hours? Write a rule to find how many times a person will blink for any number of hours.

______

4. The moon rises about an hour later each day. Suppose the moon rises at 6:45 P.M. on Monday.

a. Make a table showing the times the moon will rise for a whole week.

What pattern do you see? ______

b. What time will the moon rise on the next two Mondays?

______

c. How many days will go by before the moon rises again at 6:45 P.M.?

______

5. Katrina works at a computer during the day. Her doctor says she needs to do eye exercises. Katrina must look at an object in the distance, then at an object close to her. She must do this 5 times every 30 minutes. Make a table that shows how many times Katrina needs to do the exercises in 8 hours. Explain the pattern.

______

______

Application

COOPERATIVE LEARNING

Work with a partner to solve.

6. A clock strikes the number of the hour on the hour. It also strikes once for every half hour. For example, from 2 P.M. to 3 P.M. the clock will strike four times: once for 2:30 and three times for 3 o'clock. How many times will a clock strike in 24 hours? Describe how you solve the problem and any patterns you find.

______

______

PATTERNS BASED ON DAYS, WEEKS, AND MONTHS

Calendars have many patterns.

Example 1

If March 1 is a Monday, then March 8, March 15, and March 22 are Mondays. There are seven days in one week. If you start at 1 and count by sevens, you get Mondays.

Example 2

If you add 84 days to today's date, will it be the same day of the week?

There are 7 days in a week. 84 is a multiple of 7. So, it will be the same day of the week.

Example 3

March 1, 1993, was a Monday. What day was March 1, 1994?

To answer this problem, you need to know that there are 365 days in a year. Divide 365 by 7.

$$\begin{array}{r} 52 \text{ R1} \\ 7\overline{)365} \\ \underline{35} \\ 15 \\ \underline{14} \\ 1 \end{array}$$

If 365 is divisible by 7, then March 1, 1994, will be a Monday. Because there is a remainder of 1, you add 1 more day. So, March 1, 1994, was a Tuesday.

Reminder

60 minutes	=	1 hour
24 hours	=	1 day
7 days	=	1 week
365 days	=	1 year
366 days	=	1 leap year (every 4 years)

For some of the problems in this lesson, you may need the information on the left.

Guided Practice

1. Earl works five days a week, 48 weeks every year. He spends 30 minutes commuting to work in the morning and 30 minutes commuting home at night.

Find how many hours he spends commuting in:

a. one day (30 min + 30 min = 60 min). ___1 hour___

b. one week (1 hour per day × 5 days). ___5 hours___

c. two weeks (5 hours per week × 2 weeks). __________

d. four weeks. __________

e. 48 weeks. __________

Exercises

2. Suppose July 4 is a Saturday one year. On what day of the week would July 1 fall the next year, if the next year is not a leap year?

____

3. Every year that June 1 lands on a Saturday, Velma Johnson organizes a family reunion. Suppose the reunion falls in a leap year. How many years must she wait to have the next reunion?

____

4. Suppose you could choose to save money according to the options below. How much more money would you save with Option A than with Option B for the following lengths of time:

a. 1 day. __________

b. 10 days. __________

c. 10 weeks. __________

d. 40 weeks. __________

Option A

$2 per day

Option B

$1.50 per day

Application

Solve this problem with a partner.

5. A recent survey showed the average American watches 3 hours of television every day. If this is true for you, how many years of television will you watch by the time you are 40 years old? Explain. (Do not include leap years.)

____

____

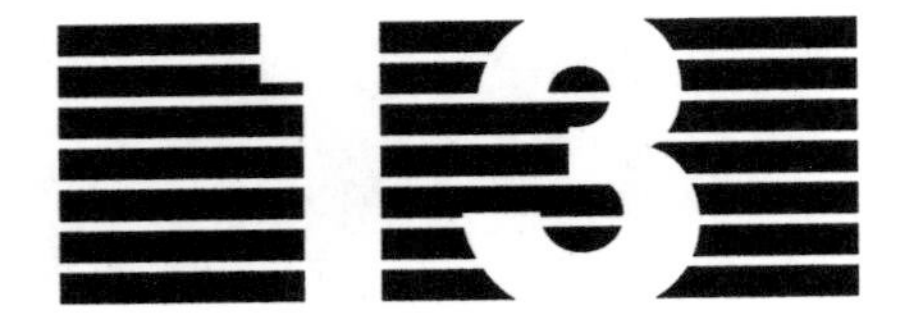

FUNCTIONS

Vocabulary

domain: the first set of numbers in a function

function: a relationship between two number sets, the domain and the range, that assigns to each number of the domain exactly one number of the range

range: the second set of numbers in a function

Helga and Fritz work at a German restaurant after school. Helga assists the manager and earns \$8 per hour. Fritz is a waiter and earns \$4 per hour and any tips he receives.

The diagrams below map the relationships between amounts of money Helga and Fritz earned (**range**) for different numbers of hours they worked (**domain**).

Helga's salary

Hours (domain)		Earnings (range)
2	→	\$16
3	→	\$24
4	→	\$32
5	→	\$40
6	→	\$48
7	→	\$56
8	→	\$64

Fritz's salary

Hours (domain)		Earnings (range)
3	→	\$20
4	→	\$24
5	→	\$29
6	→	\$31
6	→	\$32
7	→	\$37
8	→	\$39

Helga's salary can be called a **function**. For every number of hours she works, there is only *one* corresponding earning. In other words, for each domain value there is *one and only one* orange value.

Fritz's salary is not a function. He has two different earnings for working 6 hours. Because of different tips, for one 6-hour period he earned \$31 and for the other he earned \$32. So, there are *two* range values for one domain value.

Functions often show a number pattern as a rule or an equation. The equation variables are the range and domain values.

$e = 8 \times h$	The equation at the left is the function rule for Helga's salary. The variable e stands for the earnings (range), the variable h stands for hours (domain).
$e(h) = 8 \times h$	To write the function, use the notation shown at the left. Read $e(h)$ as "e of h."

You can make a table for the function. To get the different earnings, we substitute values for h into the function to compute e values.

So, $e(2) = 8 \times 2 = 16$

and $e(3) = 8 \times 3 = 24$

and so on.

Domain Hours h	Range Earnings $8 \times h$
2	16
3	24
4	32
5	40
6	48
7	56
8	64

Functions can be written as rules. You can make a table for a rule.

Reminder

$2x$ means $2 \times x$.

Function rule: $f(x) = 2x + 5$

Domain x	Range $f(x)$ $2x + 5$	
1	7	◄ $2 \times 1 + 5 = 7$
2	9	◄ $2 \times 2 + 5 = 9$
3	11	
4	13	

Function rule: $f(x) = 3x - 4$

Domain x	Range $f(x)$ $3x - 4$	
1	−1	◄ $3 \times 1 - 4 = -1$
2	2	◄ $3 \times 2 - 4 = 2$
3	5	
4	8	

Guided Practice

Relation A

Domain → Range

- a → 18
- b → 24
- c → 30
- d → 30
- e → 34
- f → 43

Relation B

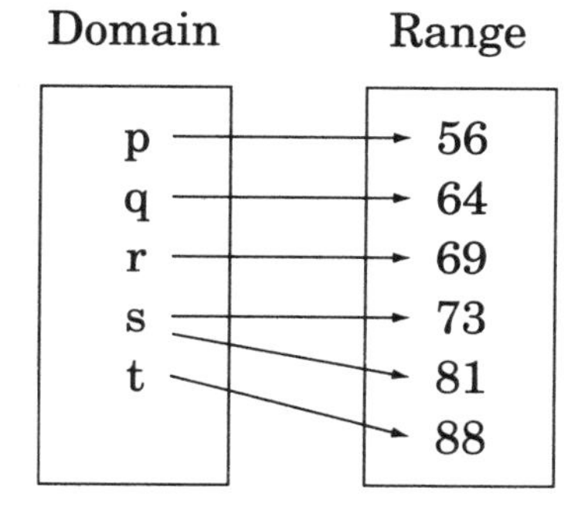

1. Look at Relation A.

a. Is there more than one number matched to any letter? no

b. Is Relation A a function? ________

2. Look at Relation B.

a. Is there more than one number matched to any letter? yes

b. Is Relation B a function? ________

3. Use the function rule $f(x) = 2x + 3$.

a. If $x = 1$, then $f(1) =$ $2 \times 1 + 3 = 5$.

b. If $x = 5$, then $f(5) =$ ________.

4. Use the function rule $f(x) = 3x + 10$.

a. If $x = 2$, then $f(2) =$ $3 \times 2 + 10 = 16$.

b. If $x = 4$, then $f(4) =$ ________.

Exercises

Look at the sets of numbers. Determine if the relationship is a function. Write *yes* or *no* and explain why.

5. Domain  Range

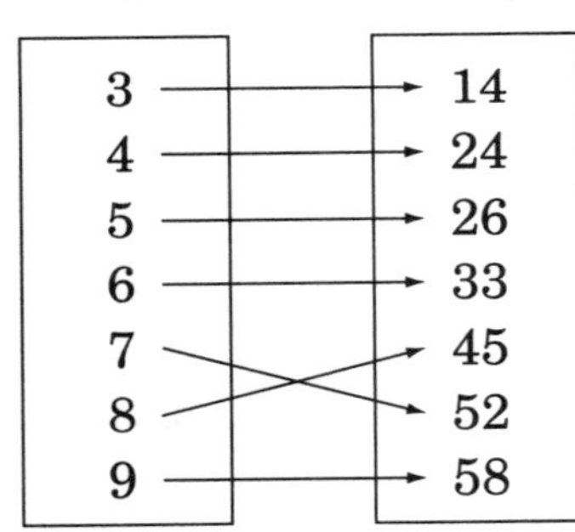

____

____

6. Domain Range

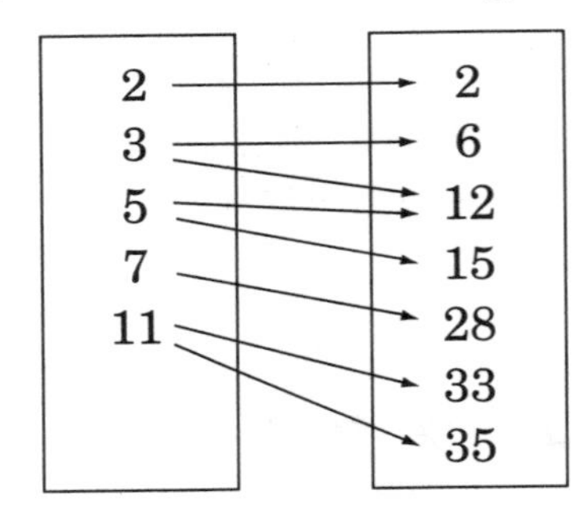

____

____

Use the given function to complete each table. Start with the given values in each table.

7. $f(x) = 2x + 7$

x	$2x + 7$
1	______
2	______
______	______
______	______

8. $f(x) = 3x - 6$

x	$3x - 6$
4	______
3	______
______	______
______	______

9. $f(x) = -2x + 5$

x	$-2x + 5$
0	______
1	______
______	______
______	______

Application

10. Fritz discovered a pattern when he was setting up square tables in a row for a party. He noticed that four people could sit at one square table. If he put two square tables together, six people could sit. With three square tables in a row, eight people could sit.

a. Cut out 12 1-centimeter squares to represent the square tables. Use your squares to discover the pattern.

b. Complete the function table to help you find the rule.

Number of Tables	Number of People
1	4
2	6
3	8
______	______
______	______
______	______

c. Write the function rule. ____________

d. Use your rule to find how many people can sit around 12 square tables set in a row. ____________

FUNCTIONS AS A SET OF ORDERED PAIRS

Vocabulary

ordered pair: a pair of numbers that gives the location of a point on a grid

When you use a function rule, you are making pairs of numbers. For example, the rule $y = x + 3$ can be used to make the function table on the right. Next to the function table in parentheses are pairs of numbers. The domain, the x value, is first. The range, the y value, is second.

Domain x	Range y $(x + 3)$	Ordered Pair (x,y)
−5	−2	(−5,−2)
−3	0	(−3,0)
−1	2	(−1,2)
1	4	(1,4)

These pairs of numbers are called **ordered pairs**. French philosopher René Descartes (1596–1650) pictured ordered pairs of numbers as points on a grid.

Look at the grid on the right. The thick horizontal number line is called the x-axis. The thick vertical number line is called the y-axis.

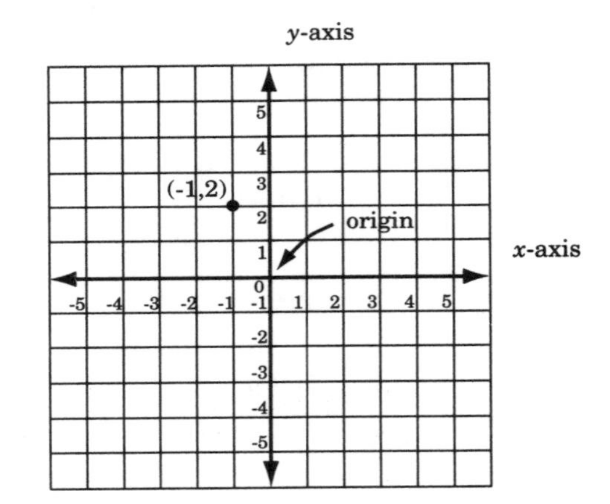

To locate the point (−1, 2), start at the center point, or origin. Move *1 unit left* and *2 units up*.

Guided Practice

1. Plot three more points for $y = x + 3$ on the grid above.

   a. (1, 4)

   b. (−3, 0)

   c. (−5, −2)

2. If x is 2, what would y be? Write these numbers as an ordered pair. (2, ____)

3. Locate and plot the point for the ordered pair in Number 2.

4. What do you notice about the points for the function $y = x + 3$?

____

Exercises

Use the function rule to complete the table.

5. $y = 2x - 3$

6. Write the ordered pair.

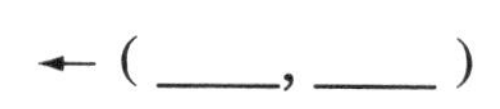

x	$2x - 3$	
-2	____	← (____, ____)
-1	____	← (____, ____)
0	____	← (____, ____)
1	____	← (____, ____)
2	____	← (____, ____)
3	____	← (____, ____)

7. Plot the points on the grid.

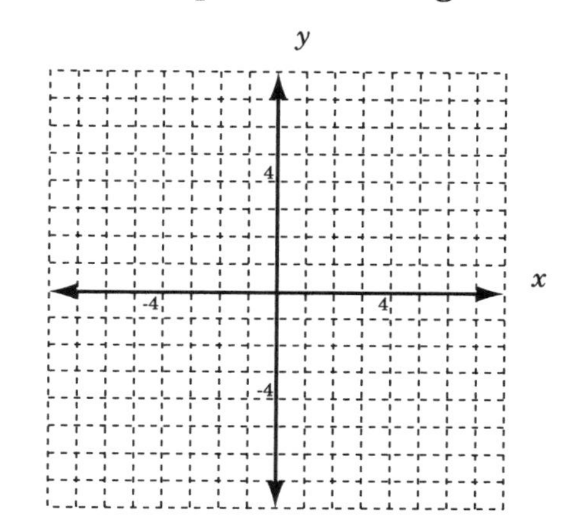

8. Look for a pattern for the points on the grid. Draw the next two points as you look at the grid from left to right. Describe how you located the points.

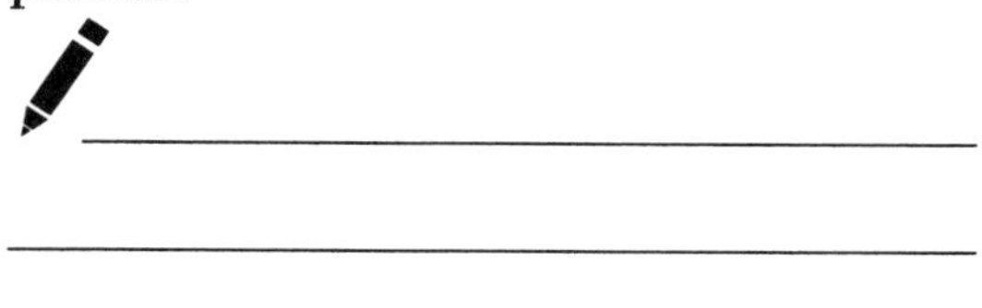

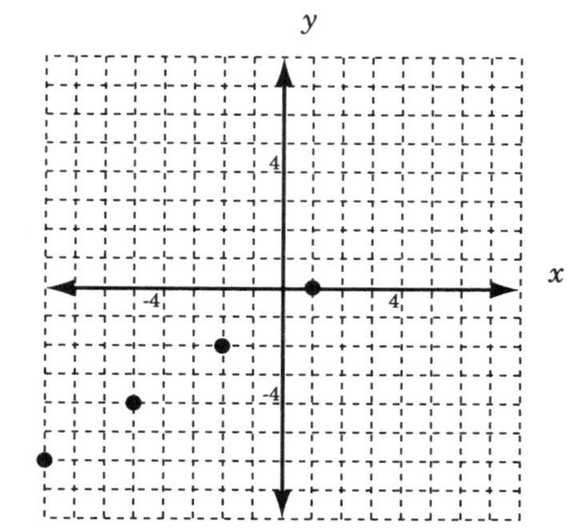

DETERMINING WHETHER A GRAPH IS A FUNCTION

You can tell by looking at a graph if it represents a function.

Example 1

This graph tracked the rock videos sold by the Funcshuns for the first eight months of 1994. The time is the domain. The number of videos sold is the range. If you choose any month, there is only one sales number.

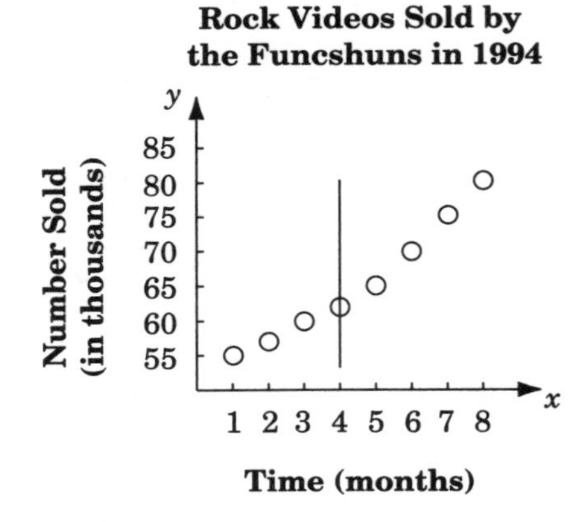

You can use a ruler (straight-edge) to test whether a graph is a function. The test is called the *vertical line test*. A graph is a function if you place the ruler anywhere *vertically* on the graph and only one point on the graph lies on the vertical line. You can slide the ruler back and forth to test other locations.

Example 2

This graph, "Shoe Size and Height," is the result of data collected from 20 student athletes comparing their height to their shoe size. The shoe size is the domain. The height is the range.

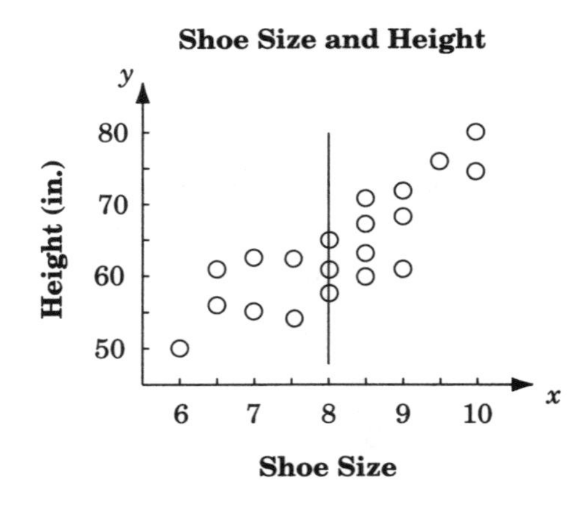

Guided Practice

1. Use the shoe size and height graph.

a. If someone asks you what the height of a person is with a shoe size of 8, how many answers can you give? ____3____

b. Use the vertical line test. Does this graph represent a function? Why or why not?

____

Exercises

Use the vertical line test to tell if the graph is a function. Write *function* or *not a function*.

2.

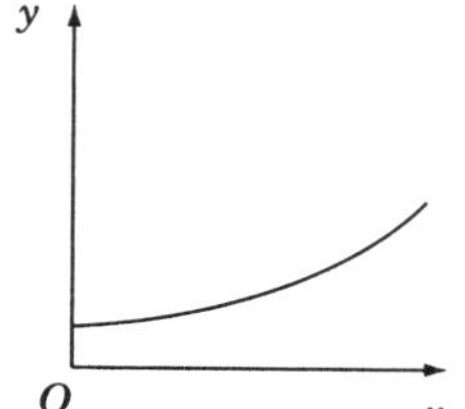

____

3.

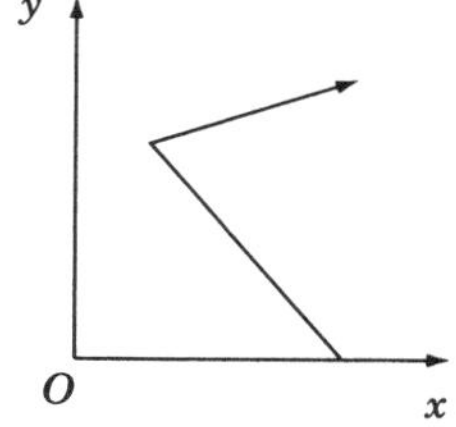

____

4.

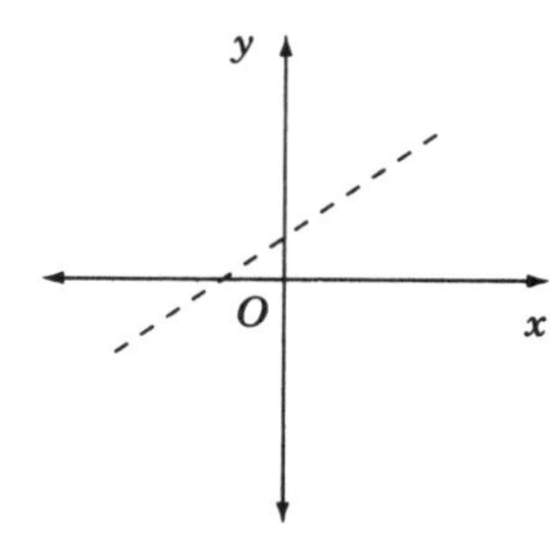

____

5.

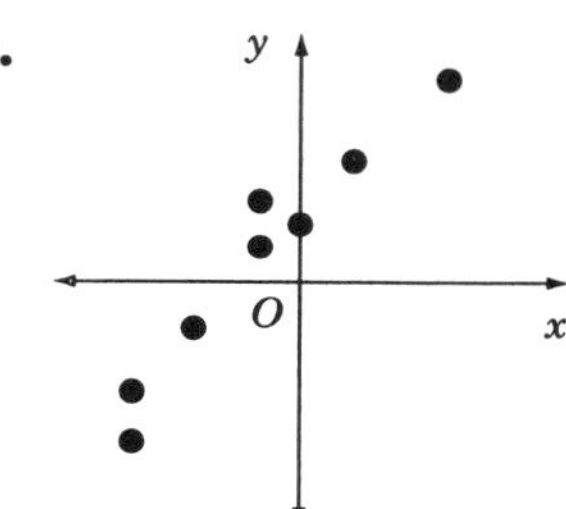

____

6.

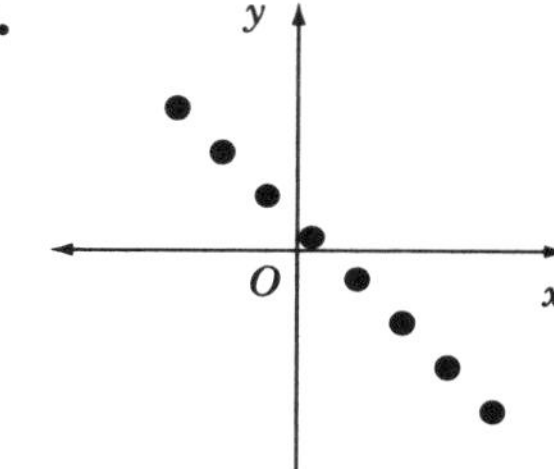

____

7.

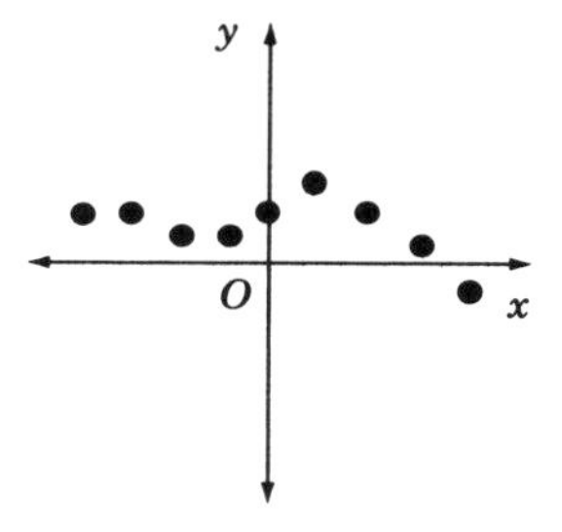

____

Application

8. Enrique was unsure if the equation $y = x$ is a function. Make a function table and graph the equation. Then explain why the equation is or is not a function.

____

____

____

GRAPHING LINEAR FUNCTIONS

Vocabulary

linear function: a function that forms a straight line when graphed

The function $y = x$ was used to compute the function table below.

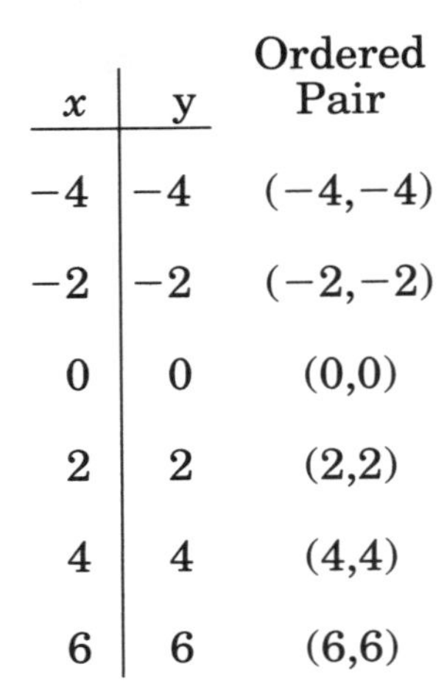

x	y	Ordered Pair
-4	-4	$(-4,-4)$
-2	-2	$(-2,-2)$
0	0	(0,0)
2	2	(2,2)
4	4	(4,4)
6	6	(6,6)

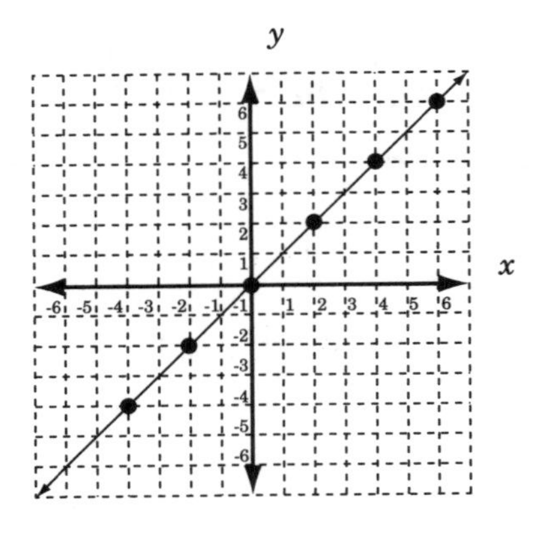

When you plot ordered pairs, the points form along a straight line. This function is called a **linear function**.

Guided Practice

1. Plot the function $y = 2x$ on the grid below.

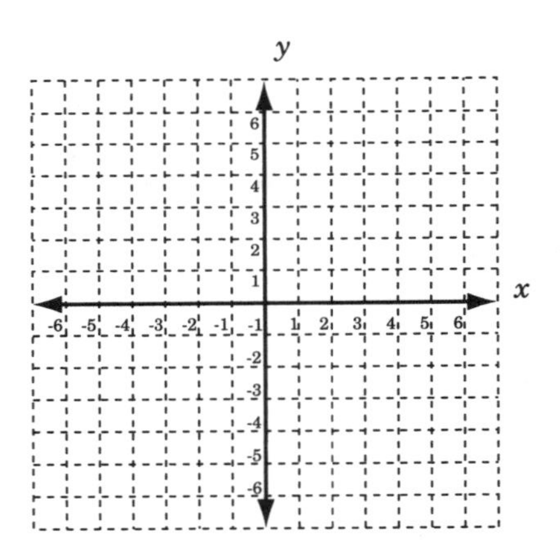

a. When $x = 0$, $y =$ ____0____.

b. When $x = 1$, $y =$ ________.

c. When $x = 2$, $y =$ ________.

d. When $x = 3$, $y =$ ________.

e. When $x = 4$, $y =$ ________.

f. When $x = 5$, $y =$ ________.

Exercises

Use the grid below to graph each linear function. Set up a function table to create ordered pairs. Plot at least six points for each graph. Draw the line connecting the points. Label each graph.

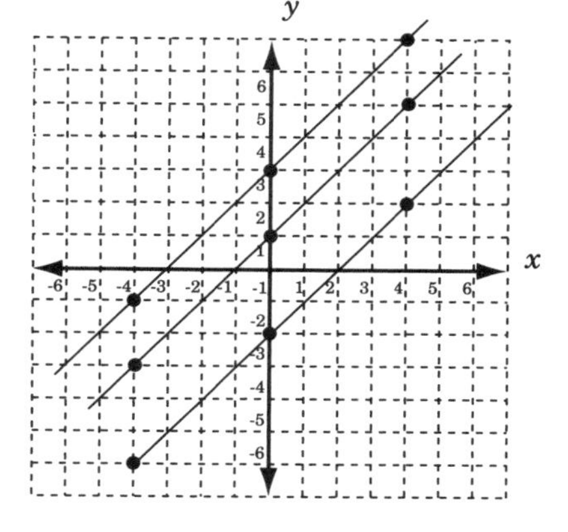

2. $y = x + 1$

3. $y = x + 3$

4. $y = x - 2$

5. Compare the graphs of the functions.

a. How are they the same?

____

____

b. How are they different?

____

c. What do you think causes these differences?

____

Application

6. Suppose you were going to graph the linear equation $y = -2x + 1$. Discuss with a partner where you think the line should go on the grid. Then plot the actual line on the grid below to check your estimate. What did you learn?

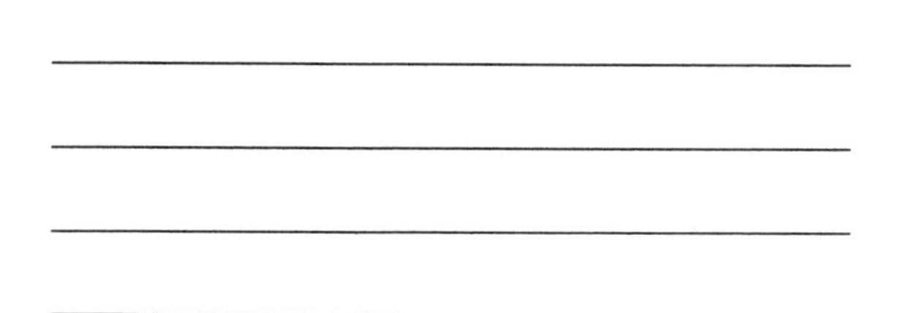

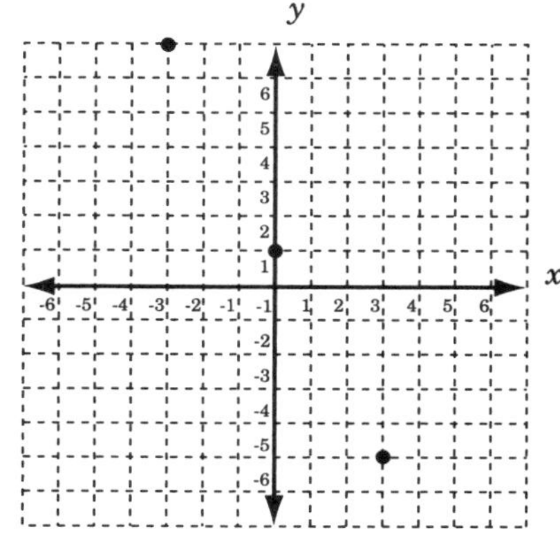

ROTATIONS

Vocabulary

rotation: movement that turns a figure about a fixed point

Video games that you play at an arcade often use **rotations**. A rotation is a movement that turns a figure. The figures below show clockwise rotations through three different angles about the center of the figure.

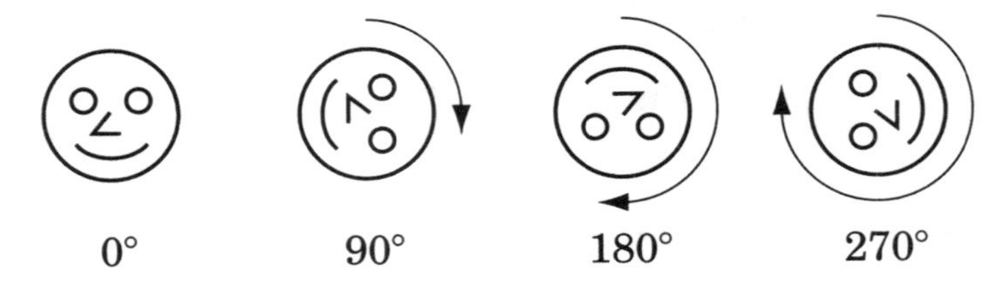

Rotations can be shown by changing the position of the figure on a grid.

Guided Practice

Reminder

Vertices are the corner points of a figure.

1. The grid below shows triangle ABC that has been rotated clockwise around point A. Find the ordered pairs for the vertices A, B, and C.

   A (<u>2</u>, <u>2</u>)

   B (____, ____)

   C (____, ____)

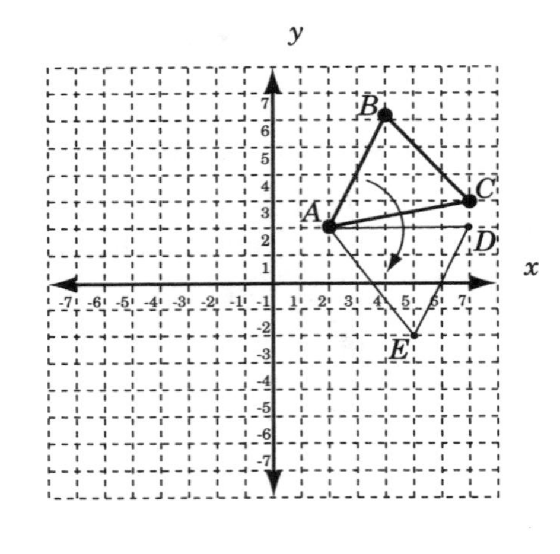

2. Find the ordered pairs for the rotated triangle.

   A (____, ____)

   D (____, ____)

   E (____, ____)

Exercises

3. Trace and cut out a triangle the same shape as the one shown on the grid that follows. Label the vertices A, B, and C. Place the cutout triangle on top of the one on the grid. Put a pencil on the cutout triangle where the origin is. Rotate the cutout triangle clockwise by the given number of

degrees and record the ordered pairs for each vertex in the table.

Rotation	A	B	C
0°	(0, 5)	(3, −4)	(−2, −1)
90°	(5, 0)		
180°			
270°			

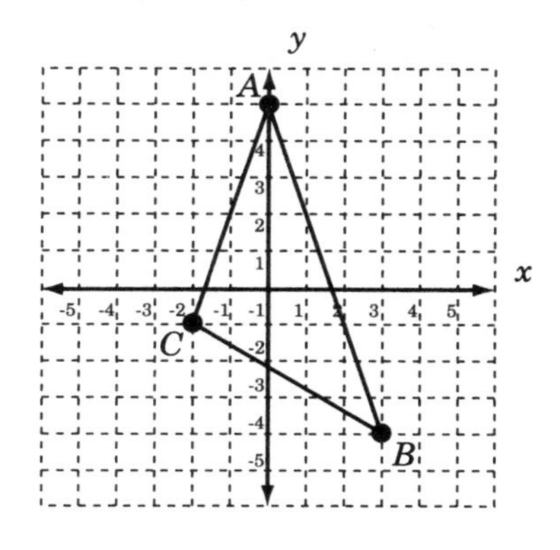

Application

4. Make a grid like the one above. Draw a triangle and label its vertices. Trace and cut out a triangle of the same shape and label the vertices. Rotate the triangle clockwise about a vertex from the original triangle. Complete the table by rotating the traced triangle 0°, 90°, 180°, and 270°. Record the ordered pairs for each rotation in the table below.

	Rotation	A	B	C
a.	0°	(____, ____)	(____, ____)	(____, ____)
b.	90°	(____, ____)	(____, ____)	(____, ____)
c.	180°	(____, ____)	(____, ____)	(____, ____)
d.	270°	(____, ____)	(____, ____)	(____, ____)

SLIDES

Vocabulary

slide: a movement of a figure in which every point moves the same distance and in the same direction

Many patterns in art from various cultures are created by slides. A **slide** is a movement of a figure in which every point moves the same distance and in the same direction.

Triangle ABC is the original position of a high school marching band. The band moves directly to the right. Triangle DEF is the new position of the band. How far to the right did the band move?

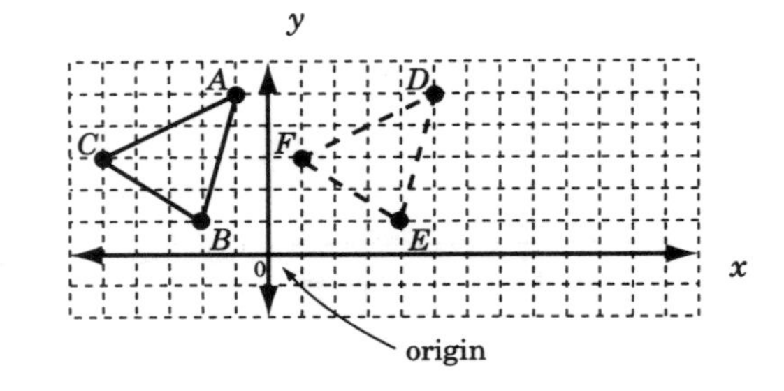

The ordered pair for A is $(-1, 5)$ and the ordered pair for D is $(5, 5)$. The difference between the two points is 6 units.

The band moved 6 units to the right.

Guided Practice

1. The band slides to the right the same distance again. The band forms a new triangle GHI. Find the ordered pair for vertex G.

   a. The ordered pair for vertex D is (__5__ , _____).

   b. The band moved how many units to the right?

   ____6____

   c. Add the number of units the band moved to the x value for vertex D.

   G(5 + _____)

   d. The ordered pair for vertex D is (_____ , _____).

Exercises

2. Look at the slides on the right.

   a. Describe slide 1.

   ____

   ____

   b. Describe slide 2.

   ____

   ____

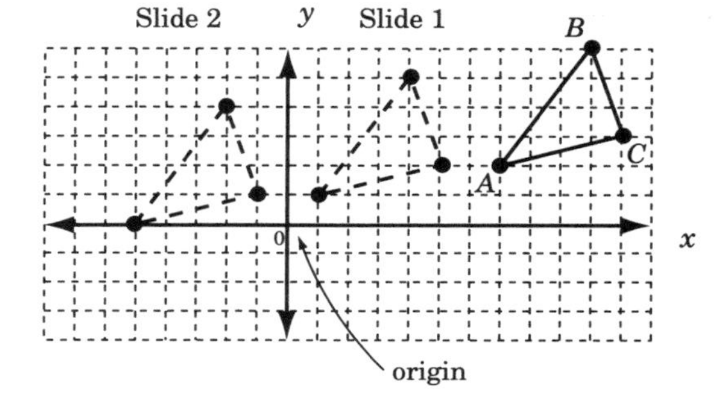

Complete the table below for each vertex of slides 1 and 2 above.

	Figure	A	B	C
3.	Triangle ABC	(____ , ____)	(____ , ____)	(____ , ____)
4.	Slide 1	(____ , ____)	(____ , ____)	(____ , ____)
5.	Slide 2	(____ , ____)	(____ , ____)	(____ , ____)

Application

6. Use the rectangle at the right. Make two slides by moving the rectangle 4 units to the right and 2 units down for each slide. Describe the strategy you used to draw the new rectangle.

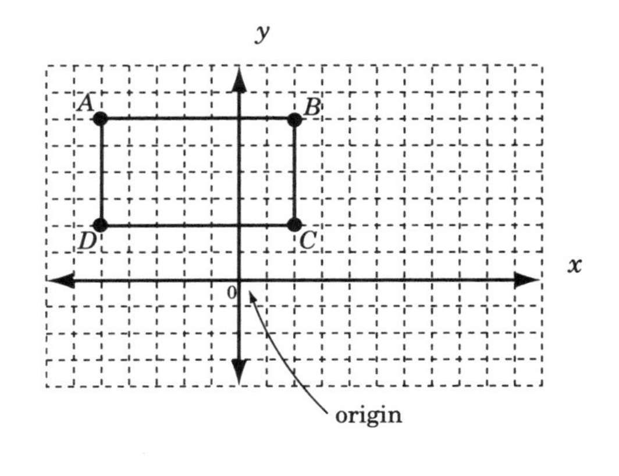

____

____

____

____

____

____

COMBINING ROTATIONS AND SLIDES

On Jasmine's new video game, Geogrids, she must move triangle ABC 180° counterclockwise onto triangle DEF.

Jasmine does this by combining a slide and a rotation as shown on the grid.

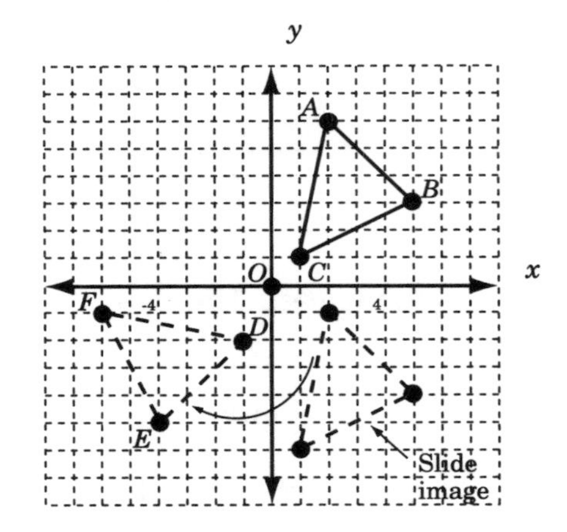

First, she slides triangle ABC 7 units down. She discovers that for a downward vertical slide the values for y decrease by 7 and the values for x stay the same.

$(x, y) \rightarrow y - 7)$

Then, Jasmine rotates the slide image 90° clockwise around the origin O. She discovers a rule for a 90° clockwise rotation around the origin.

$(x, y) \rightarrow (y, -x)$

Guided Practice

1. Trace the original triangle ABC above. Label the vertices of the new triangle $A'B'C'$. Place the tracing over the triangle ABC. Pin the paper down at the origin O and rotate the triangle 90° clockwise around the origin.

   a. What are the coordinates for vertices A, B, and C?

   b. What are the new coordinates for vertices A', B', and C'?

   A'(6 , −2), B'(____, ____), C'(____, ____)

2. How can you move triangle $A'B'C'$ onto triangle DEF in the above?

____

Exercises

Look at triangle *ABC* on each grid. Describe two ways you can move it to match it onto triangle *DEF*. Tracing paper may help you find the movements.

3.

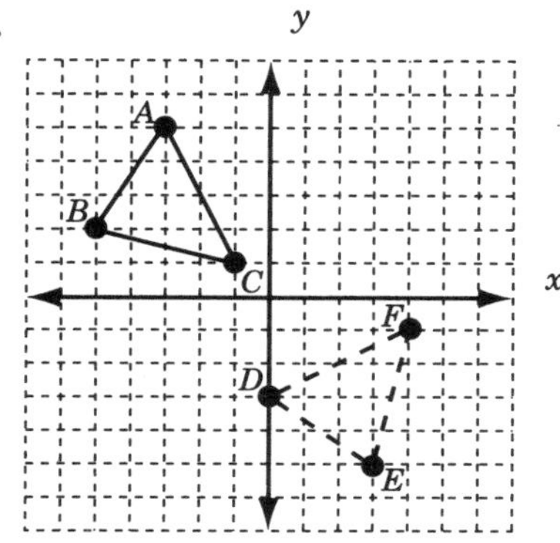

a. 1st way: ____________________

b. 2nd way: ____________________

4.

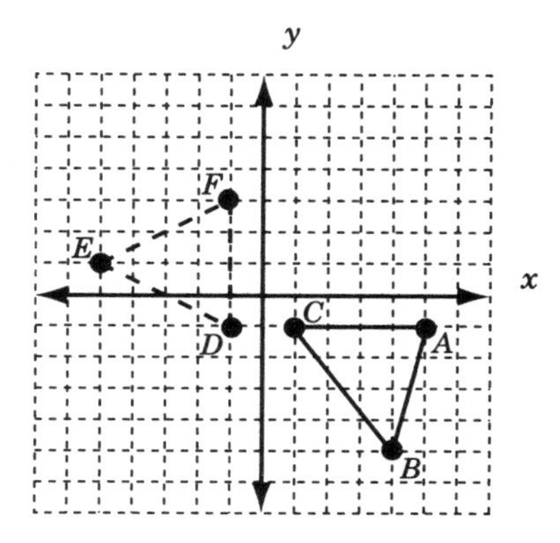

a. 1st way: ____________________

b. 2nd way: ____________________

Application

Work with a partner.

5. Make a grid like the one above. Draw triangle *ABC* and label its vertices. Trace the original triangle *ABC* and label the vertices. Without your partner watching, find a new location for the triangle using one slide and one rotation in any order. Record your movements and mark the location of new triangle *XYZ*.

Your partner must now describe how your triangle was moved. After your partner tries, change roles.

6. Record your observations or discoveries. ____________

7. Determine the coordinate rule for a 90° *counterclockwise* rotation.

$(x, y) \rightarrow$ (_____, _____)

PATTERNS RELATING ANGLE MEASURE IN POLYGONS

Vocabulary

measure of an angle: the degrees of rotation between two rays with a common vertex

polygon: a closed plane figure made up of sides and angles

Reminder

A *ray* is a part of a line that has one endpoint and extends without end in one direction.

The **measure of an angle** is the number of degrees of rotation between two rays with a common vertex. In the diagram on the right, the angle was formed by rotating ray *VA* 45° about vertex *V*.

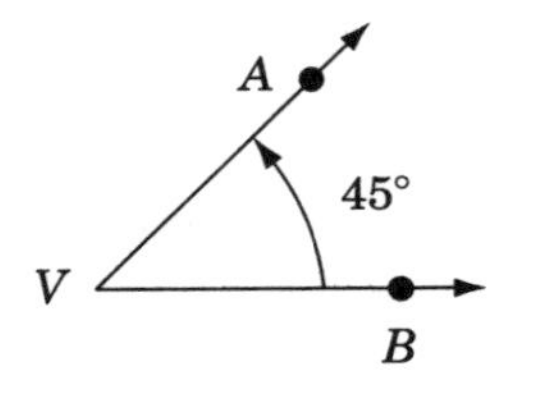

All angles form one of four types of measures.

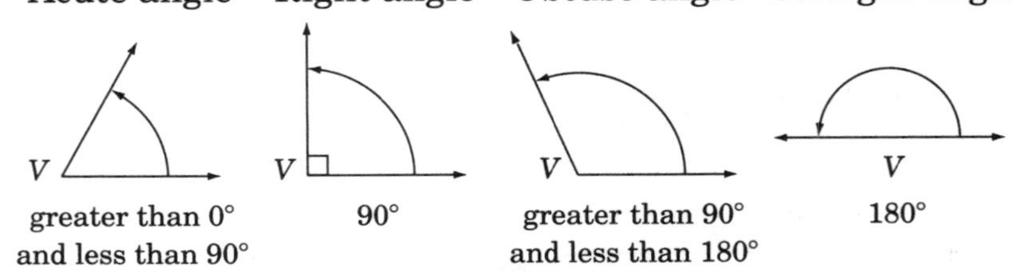

A triangle is a **polygon** with three sides and three angles. If you tear off each angle of a triangle, you can place them together and form a straight angle or line. The measure of a straight line is 180°.

So, the sum of the three angles of a triangle is 180°.

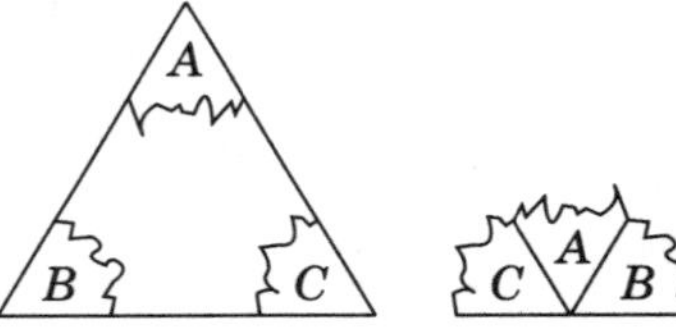

Guided Practice

1. Triangle *ABC* has angle $A = 45°$ and angle $B = 90°$.

   a. What is the sum of the two given angles?

   135°

   b. What is the sum of the measures of all the angles of the triangle? ______

   c. What is the measure of the third angle?

   ______

Exercises

The measures of two angles in a triangle are given.

a. Find the measure of the unknown angle.

b. Is the unknown angle acute, right, or obtuse? Write *acute, right,* or *obtuse.*

2.

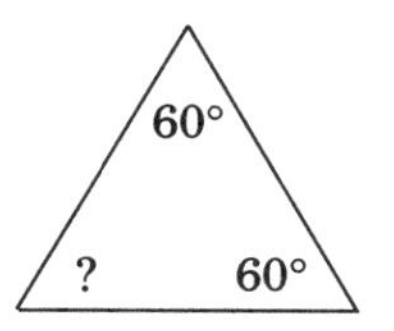

a. ____________

b. ____________

3.

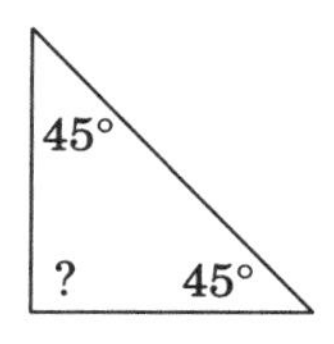

a. ____________

b. ____________

4.

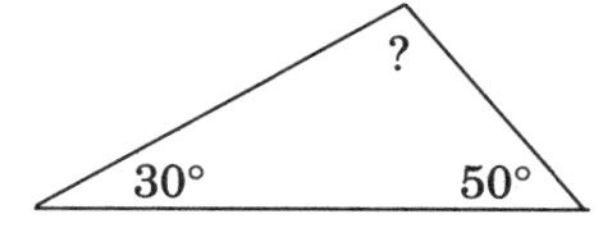

a. ____________

b. ____________

Application

5. The diagram on the right shows that exactly one diagonal can be drawn from one vertex of the four-sided polygon. Two triangles are formed. Because the sum of the measures of the angles in a triangle is 180°, the sum of the measures of a four-sided figure is $2 \times 180°$, or 360°. With a partner, explore polygons with more sides. Draw the diagonals from one vertex and complete the chart below.

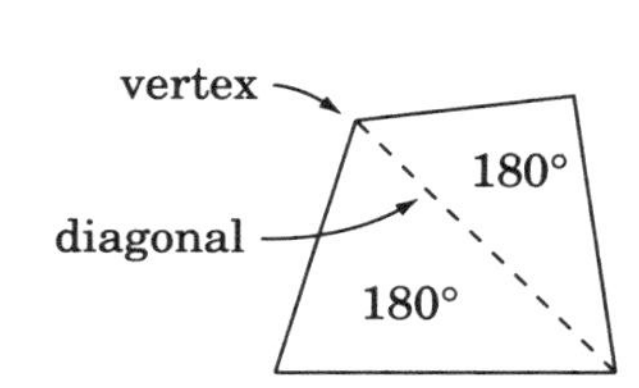

Polygons

	Number of Sides	Number of Diagonals	Number of Triangles	Sum of the Angles
	3	0	1	180°
	4	1	2	360°
a.	5			
b.	6			
c.	7			
d.	n			

6. Look for a pattern and write a rule to find the sum of the measures of the angles of a polygon with n sides.

____

____

PATTERNS RELATING LINEAR MEASUREMENTS IN AREA

Reminder

Area is measured in square units, such as square feet (ft^2).

Calvin measures the community bandstand as 10 feet by 10 feet. He calculates its area as 100 square feet. Calvin wants to expand the bandstand floor to 20 feet by 20 feet. He says that will double the floor area. Tanya thinks he is wrong. She says the floor area will be four times larger. To solve their problem, they need to know how to calculate area.

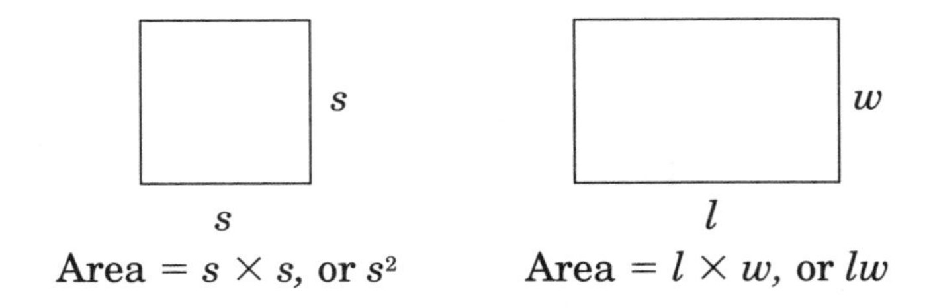

$$10\ ft \times 10\ ft = 100\ ft^2$$

Calvin is right about the area of the bandstand.

However, doubling the lengths of the sides of a square does not double the area. If Calvin draws a diagram like the one below, he will see that the area increases four times. Tanya is right about the new area.

$$20\ ft \times 20\ ft = 400\ ft^2$$

100 ft²

10 ft (side), 10 ft (bottom)

100 ft²	100 ft²
100 ft²	100 ft²

20 ft (side), 20 ft (bottom)

Guided Practice

Reminder

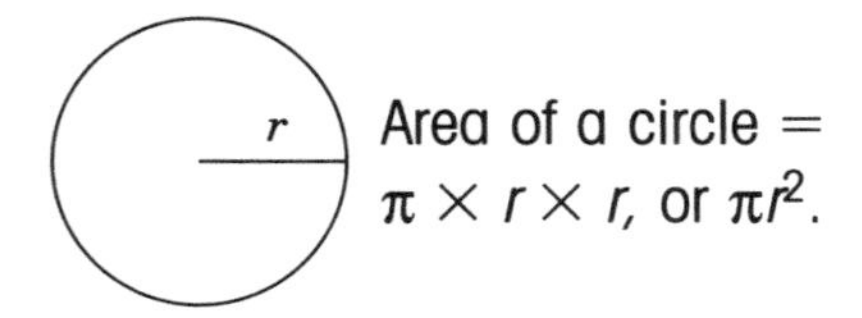

1. A circle has a radius of 2 cm. What happens to the area if the radius is tripled?

a. Write the formula for finding the area of a circle.

$A = \pi \times r \times r$, or πr^2

b. Substitute 2 for the radius and 3.14 for π.

____

c. Calculate the area of the circle. ____________

d. Triple the radius and find the area.
$r =$ ____________

e. How many times greater is the new area?

____

Exercises

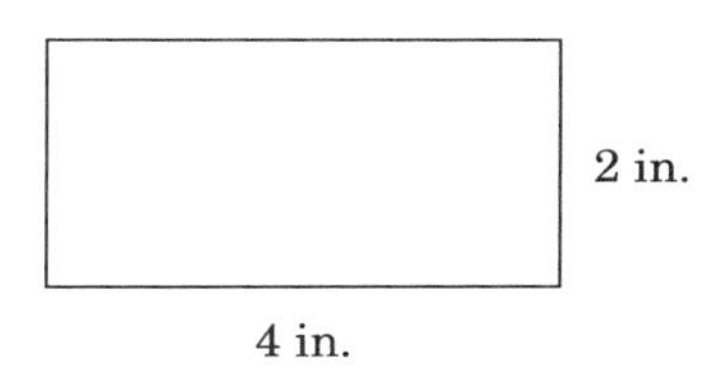

2. The length of a rectangle is 4 inches. The width is 2 inches. The area is 8 square inches. Find the area when the length is increased by the given factor. Complete the table below.

Factor	2	3	4	5
New Length (in.)	**a.** 8	**c.**	**e.**	**g.**
Area (in.^{2})	**b.**	**d.**	**f.**	**h.**

3. Describe the pattern.

____

Application

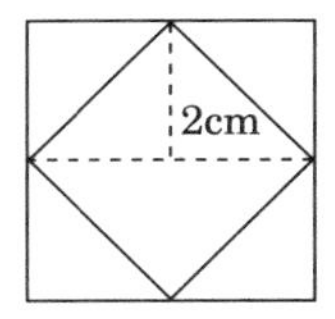

4. Use the formula for the area of a triangle ($A = \frac{1}{2}bh$) to find the area of the small square. ____________

a. Find the areas when the height is increased by the given factors.

Factor		2	3
Height	2		
Base	4		
Area (small square)	8		
Area (large square)	16		

b. Describe the pattern. ________________________

PYTHAGOREAN TRIPLES

Vocabulary

Pythagorean theorem: In all right triangles, the square of the hypotenuse is equal to the sum of the squares of the legs

Pythagorean triple: Three whole numbers where the sum of the squares of the smaller integers is equal to the square of the greater integer

The **Pythagorean theorem** is named after a Greek mathematician named Pythagoras (570 – 500 B.C.). More than 500 years earlier, the Egyptians were using the idea to make right angles for surveying. They used a rope with twelve evenly spaced knots. The diagram on the right shows how they arranged the rope to form the right angle.

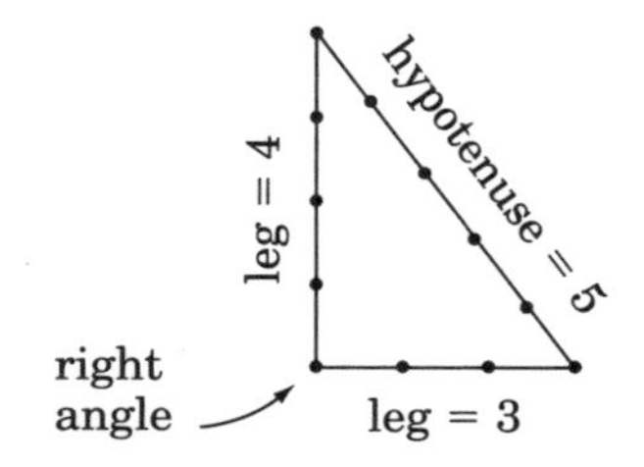

The sides of this triangle are 3-4-5. This is called a **Pythagorean triple**. The Pythagorean theorem states that for all right triangles, the square of the hypotenuse is equal to the sum of the squares of the legs. The diagram on the right shows this theorem.

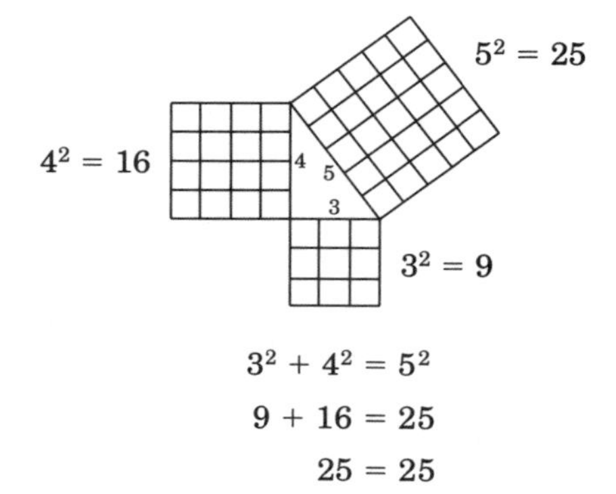

You can check to see if a set of three whole numbers is a Pythagorean triple. Let a and b be the lengths of the legs and let c be the length of the hypotenuse.

$$a^2 + b^2 = c^2$$

4, 6, 9	6, 8, 10
$4^2 + 6^2 \ ? \ 9^2$	$6^2 + 8^2 \ ? \ 10^2$
$16 + 36 \neq 81$	$36 + 64 = 100$
This is not a Pythagorean triple.	This is a Pythagorean triple.

Reminder

a^2 means $a \times a$.

Guided Practice

1. Are the numbers 5, 6, and 8 a Pythagorean triple?

a. 5^2 = 25 6^2 = 36 8^2 = ______

b. The sum of 5^2 and 6^2 is ______.

c. Are the numbers 5, 6, and 8 a Pythagorean triple? ______

Exercises

If a set of three whole numbers is a Pythagorean triple, then these numbers could be lengths of a right triangle. Test each set of numbers to see if they could be sides of a right triangle. Write *yes* or *no*. Use a calculator.

2. 13, 15, 18 ______ **3.** 6, 8, 10 ______ **4.** 8, 15, 17 ______

5. 5, 12, 13 ______ **6.** 13, 16, 21 ______ **7.** 20, 21, 29 ______

The two legs *a* and *b* of a right triangle are given. Find the hypotenuse *c*.

8. 9 and 12 ______ **9.** 16 and 30 ______ **10.** 12 and 16 ______

Application

Complete the following activity with a partner.

11. The Pythagorean triple 3-4-5 can be used to create more Pythagorean triples. The triples 6-8-10 and 9-12-15 are related to the triple 3-4-5.

a. Discuss the relationship between these triples. Record your observations.

______

______

b. Discover two more triples that are related to the 3-4-5 triple. Record your findings.

______

c. Does each triple you found obey the Pythagorean theorem? Explain your answer.

______

GOLDBACH'S FIRST CONJECTURE

Vocabulary

conjecture: a guess that something is true

Look at the equations below. What do these addends have in common? What do their sums have in common?

2 + 2 = 4	3 + 3 = 6	3 + 5 = 8
5 + 5 = 10	5 + 7 = 12	7 + 7 = 14
5 + 11 = 16	7 + 11 = 18	7 + 13 = 20

Reminder

A prime number has exactly two factors; one and itself.

All of the addends are prime numbers.

All of the sums are even numbers.

Goldbach's First Conjecture states that every even number greater than 2 can be represented as the sum of two prime numbers.

It is a **conjecture** because it is a conclusion based on observations. Christian Goldbach (1690–1764) was never able to prove that his conjecture holds true for all even numbers. But no one has been able to prove that it is not true.

Guided Practice

1. There are three pairs of primes that have a sum of 22. Find the prime addends.

a. ___3___ + ________ = 22

b. ___5___ + ________ = 22

c. ________ + ________ = 22

2. What two primes have a sum of 24?

________ and ________

Exercises

Write each even number as a sum of two primes.

3. 24 = ______ + ______ **4.** 26 = ______ + ______

5. 28 = ______ + ______ **6.** 30 = ______ + ______

7. 32 = ______ + ______ **8.** 34 = ______ + ______

9. 36 = ______ + ______ **10.** 38 = ______ + ______

11. 40 = ______ + ______ **12.** 42 = ______ + ______

13. 44 = ______ + ______ **14.** 46 = ______ + ______

Use a table of prime numbers from 1 to 100 to help you find the prime addends. Write each even number as a sum of two primes.

15. 68 = ______ + ______ **16.** 76 = ______ + ______

17. 84 = ______ + ______ **18.** 96 = ______ + ______

19. 100 = ______ + ______ **20.** 110 = ______ + ______

Application

21. Work with a partner. Look back at all the even numbers presented in this lesson. Only one even number used the addend 2.

a. What was this even number? ______

b. Explain why this is the only even number that could use the number 2 as an addend.

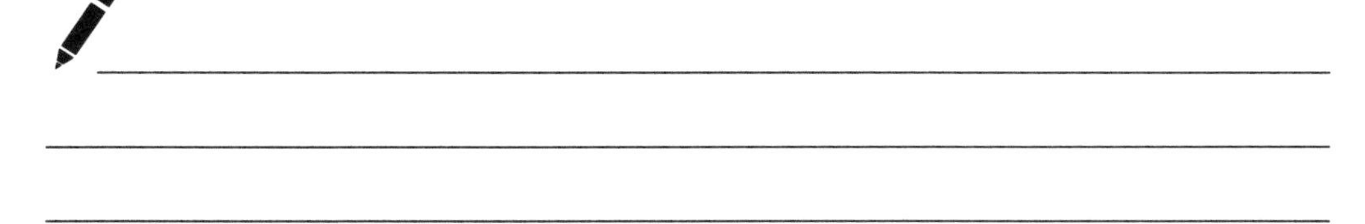

24 GOLDBACH'S SECOND CONJECTURE

Look at the equations below. What do these addends have in common? What do their sums have in common?

$2 + 2 + 3 = 7$ $3 + 3 + 3 = 9$ $3 + 3 + 5 = 11$

$3 + 3 + 7 = 13$ $3 + 5 + 7 = 15$ $5 + 5 + 7 = 17$

$5 + 7 + 7 = 19$ $7 + 7 + 7 = 21$ $5 + 7 + 11 = 23$

All of the addends are prime numbers.

There are three addends in each sum.

All of the sums are odd numbers.

Goldbach's Second Conjecture states that every odd number greater than or equal to 7 can be represented as the sum of three prime numbers.

Guided Practice

1. Find four groups of three prime addends that have a sum of 25.

a. 3 + 3 + ___19___ = 25

b. 3 + ___5___ + 17 = 25

c. ______ + ______ + ______ = 25

d. ______ + ______ + ______ = 25

2. What three primes have a sum of 27?

______ and ______ and ______

Exercises

Write each odd number as the sum of three prime numbers.

3. 29 = ______ + ______ + ______
4. 31 = ______ + ______ + ______

5. 33 = ______ + ______ + ______
6. 35 = ______ + ______ + ______

7. 37 = ______ + ______ + ______
8. 39 = ______ + ______ + ______

9. 41 = ______ + ______ + ______
10. 43 = ______ + ______ + ______

11. 61 = ______ + ______ + ______
12. 83 = ______ + ______ + ______

Find three prime addends for each odd number. Look for patterns.

	Odd Number			
13.	45	3		
14.	57	3		
15.	65	3		
16.	87			
17.	99			
18.	103			

Application

19. Jody made a conjecture. "Every odd number greater than 7 has more than one set of three prime numbers." Explore her conjecture with your partner. Write an explanation that either supports or disproves her conjecture. (Hint: You only need one example to show that a conjecture is false.)

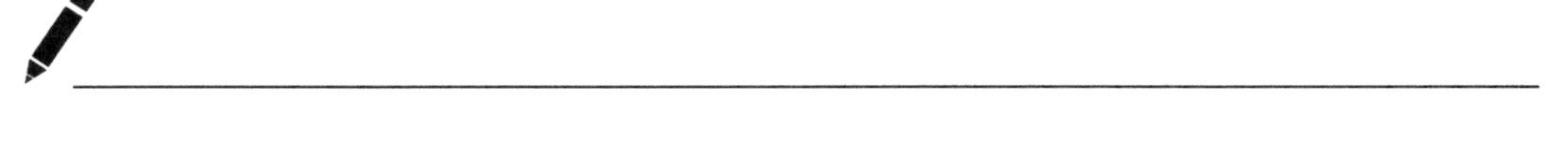

LESSONS

1-4 CUMULATIVE REVIEW

Find the rule for each number pattern. Write the next three numbers in the pattern.

1. 5, 12, 19, 26, ____, ____, ____

2. 8, 20, 32, 44, ____, ____, ____

3. 3, 12, 21, 30, ____, ____, ____

4. 1.1, 2.3, 3.5, 4.7, ____, ____, ____

Write the first five multiples for the given number.

5. 6: ______________________

6. 8: ______________________

List the next three multiples. Then find the 30th multiple.

7. 7, 14, 21, ____, ____, ____

30th multiple: __________

8. 12, 24, 36, ____, ____, ____

30th multiple: __________

Complete the table using the rule.

9. $5n + 4$

n				
$5n + 4$				

10. $7m - 4$

m				
$7m - 4$				

Write the rule for each pattern. Then find the 90th term.

11. 6, 10, 14, 18, ...

Rule: __________

90th term: ______

12. 8, 17, 26, 35, ...

Rule: __________

90th term: ______

13. 5, 12, 19, 26, ...

Rule: __________

90th term: ______

14. Mr. Alfonso is having his car fixed. The mechanic said the parts alone cost $45. Labor costs $30 per hour. Set up a table and use a pattern to show the costs from 1 hour to 6 hours.

LESSONS

5-8 CUMULATIVE REVIEW

Draw the next shape. Describe the pattern.

1.

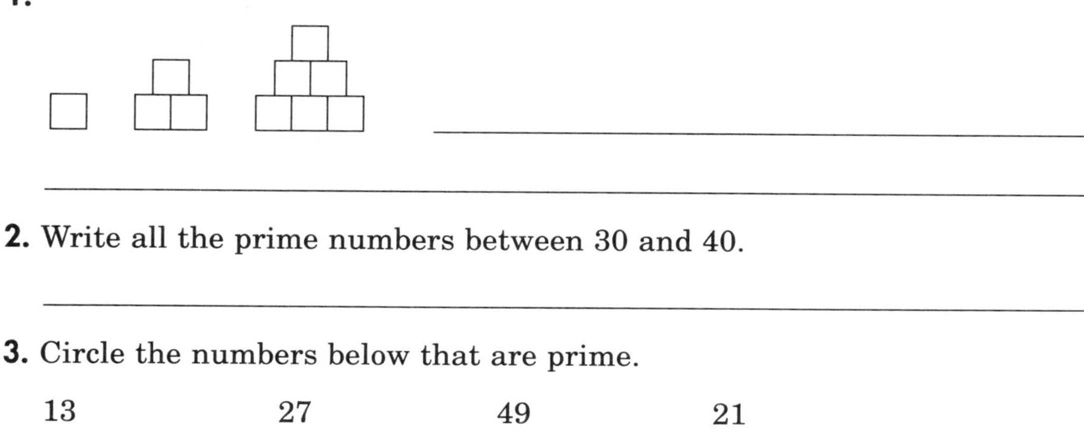

2. Write all the prime numbers between 30 and 40.

3. Circle the numbers below that are prime.

13 27 49 21

63 23 51

Use a factor tree to find the prime factorization for the number.

4. 28

5. 84

Find the next term in each pattern. Describe how you found it. You may use a calculator.

6. 2, 10, 50, 250, ________

7. 800, 400, 200, 100, ________

Solve the problem.

8. Kaylee tried a reading experiment. She had to read a 120-page book. On the first day, she read one page. On the second day, she read two pages. On the third day, she read three pages. On the fourth day, she read four pages and so on until she finished the book. On what day did she finish the book?

LESSONS

9-12 CUMULATIVE REVIEW

Write the next three terms in the sequence. Describe the pattern.

1. 3, 4, 6, 9, 13, ____, ____, ____, ...

____

2. 4, 4, 8, 12, 20, 32, ____, ____, ____, ...

____

Use the rule to write the first 6 terms in a pattern. Let *a* stand for the term number.

3. $a^2 + 5$

____, ____, ____, ____, ____, ____, ...

4. $4a + a$

____, ____, ____, ____, ____, ____, ...

Draw the next dot figure in the pattern.

5. 

How many dots are in the 20th figure? ____________

Solve each problem. You may want to make a table to help you.

6. At 3 o'clock, the hour and minute hands on a clock form a 90° angle. The hands also form a right angle a little after 3:30. How many times does this happen from 12 noon to 12 midnight?

____

7. It's 7 A.M. What time will it be in 10 hours? In 100 hours?

____

8. If July 1 was a Thursday, on what day will July 1 be on the next year? Remember, there are 365 days in a year. (Assume that the year is not a leap year.)

____

9. If you add 91 days to today's date, what day of the week will it be? Explain.

____

LESSONS

13-16 CUMULATIVE REVIEW

1. Look at the sets of numbers. Determine if the set is a function. Write yes or no.

Domain		Range
3	→	25
4	→	29
5	→	32
6	→	41
7	→	53
8	→	58
9	→	72

2. Use the given function to complete the function table.

$f(x) = 3x - 2$

x	$3x - 2$
0	____
1	____
2	____
3	____

Use the function rule to complete the table, find ordered pairs, and graph the linear function.

3. $y = 2x - 4$

x	$2x - 4$
−2	____
0	____
2	____
4	____
6	____

4. Write the ordered pairs

← (____, ____)

← (____, ____)

← (____, ____)

← (____, ____)

← (____, ____)

5.

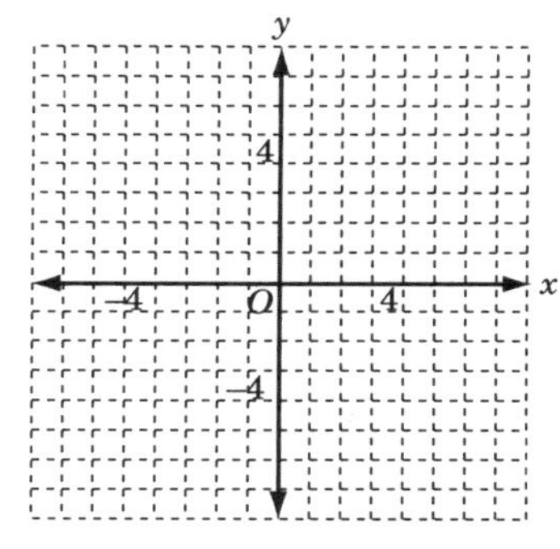

Use the vertical line test to decide if the graph is a function or not. Write *function* or *not a function*.

6.

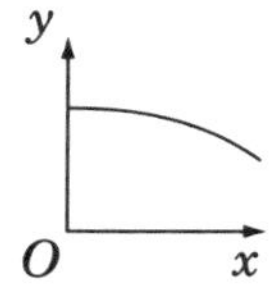

____

7.

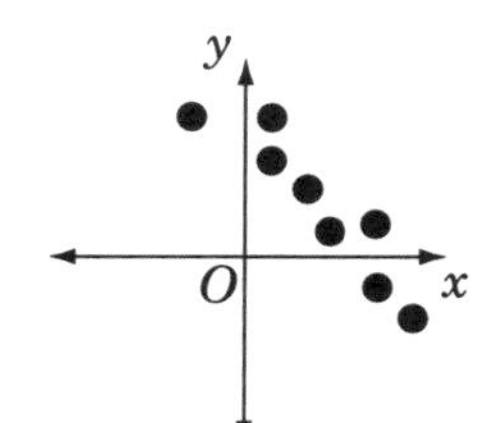

____

8.

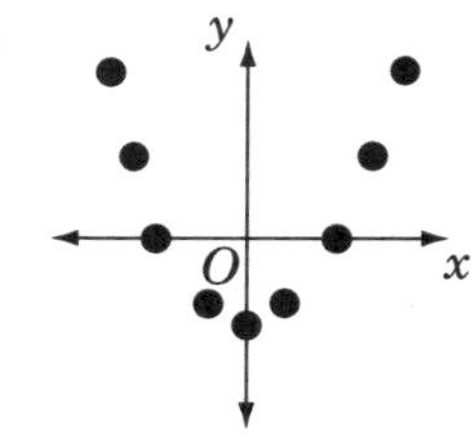

____

17-21 CUMULATIVE REVIEW

Use tracing paper to rotate and slide triangles.

1. Rotate the triangle 90° clockwise about the origin. Draw the new triangle on the grid and label it triangle *DEF*. Write the ordered pair for each vertex below.

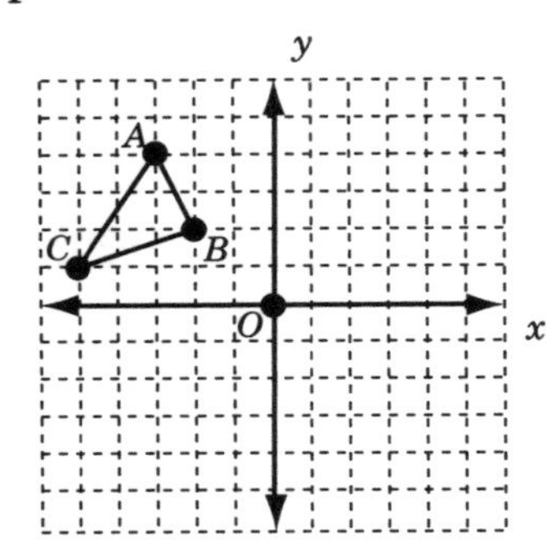

A (____, ____) *B* (____, ____)

C (____, ____) *D* (____, ____)

E (____, ____) *F* (____, ____)

2. The grid below shows a slide.

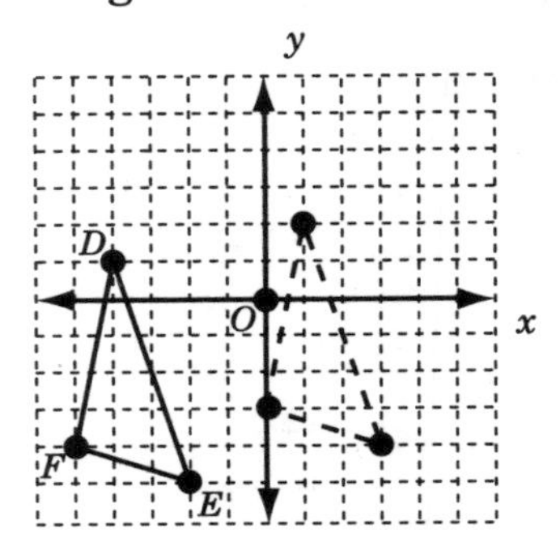

a. How did the triangle move?

____

____

b. Repeat the slide one more time.

Draw the new triangle above.

3. Look at the grid on the right. Describe how to move triangle *ABC* onto triangle *DEF*.

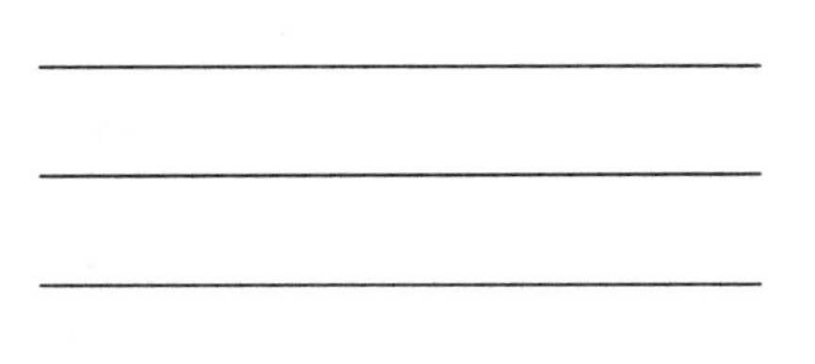

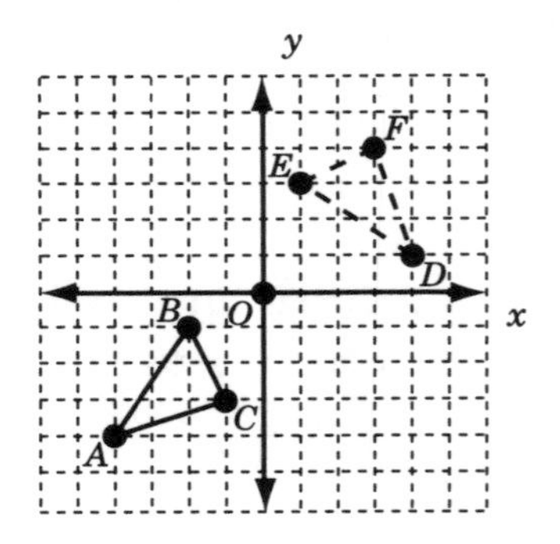

Name the sum of the angles of each polygon.

4. a three-sided polygon __________

5. four-sided polygon __________

6. five-sided polygon __________

The radius of a circle is 5 cm.

7. If the radius is doubled, by what factor is the area increased? ________

8. If the radius is tripled, by what factor is the area increased? ________

LESSONS

22-24 CUMULATIVE REVIEW

1. Look at the three statements below. Which statement is a summary of the Pythagorean theorem? (Choose *a*, *b*, or *c*.)

   a. The sum of two sides of a triangle is equal to the third side.

   b. In all right triangles, the square of the hypotenuse is equal to the sum of the squares of the legs.

   c. In a right triangle, add the legs and square them. The result is the hypotenuse squared.

Is the set of numbers a Pythagorean triple? Write *yes* or *no*.

2. 12, 15, 17 ________ **3.** 15, 20, 25 ________ **4.** 16, 30, 34 ________

The two legs of a right triangle are given. Use the Pythagorean Theorem to find the hypotenuse.

5. 12 and 16 ________ **6.** 7 and 24 ________ **7.** 24 and 45 ________

Write the letter of the statement that matches each.

8. Goldbach's First Conjecture ________

9. Goldbach's Second Conjecture ________

a. Any even number greater than 4 can be represented by the sum of two odd numbers.

b. Any odd number greater than or equal to 7 can be represented by the sum of three prime numbers.

c. Any even number greater than 2 can be represented by the sum of two prime numbers.

d. Any odd number greater than 5 can be represented by the sum of three or more odd numbers.

Use Goldbach's First Conjecture to find a set of numbers for the given sum.

10. 56 = ____________ **11.** 64 = ____________

Use Goldbach's Second Conjecture to find a set of numbers for the given sum.

12. 49 = ____________ **13.** 57 = ____________

ANSWER KEY

LESSON 1 (pp. 2 - 3)

1. a. 3 hours **b.** Add 3. **c.** 2 more
3. Add 3; 26, 29, 32. **5.** Add 5; 24, 29, 34.
7. Add 0.9; 6.1, 7.0, 7.9.
9. Add 13; 64, 77, 90.
11. Check students' work.

LESSON 2 (pp. 4 - 5)

1. a. 6 **b.** 30, 36, 42
c. 78; Answers may vary. Multiply 6 by the term number 13. $6 \times 13 = 78$.
3. 52, 65, 78; 260 5. 40, 50, 60; 200
7. 44, 55, 66; 220
9. Students could either divide $168 by 12 or complete the table and extend to $168 or more.

LESSON 3 (pp. 6 - 7)

1. a. $8h + 30$ **b.** $8 \times 2 + 30$ **c.** 46
3. 5, 8, 11, 14 **5.** 16, 25, 34, 43
7. $5\frac{1}{2}$, $11\frac{1}{2}$, $17\frac{1}{2}$, $23\frac{1}{2}$
9. a. Check students' work. **b.** $148

LESSON 4 (pp. 8 - 9)

1. a. 4 **b.** 4 **c.** $4n$ **d.** add 1 **e.** $4n + 1$
f. $4 \times 2 + 1 = 9$ **g.** $4 \times 70 + 1 = 281$
3. $10n + 1$; 801 **5.** $6n$; 480
7. $3n + 1.5$; 241.5

LESSON 5 (pp. 10 - 11)

1. a. 14 **b.** 6 **c.** 20
3. Each figure is divided evenly by the term number.
5. Half of the previous shaded shape is shaded.

LESSON 6 (pp. 12 - 13)

1. 2 **3.** Composite
5. 31; The only factors of 31 are 1 and 31.
7. False **9.** False
11. Check students' tables.

LESSON 7 (pp. 14 - 15)

1. a. 8 **b.** 4 **c.** 2 **d.** $2 \times 2 \times 2 \times 5 = 40$
3. $30 = 2 \times 3 \times 5$ **5.** $150 = 2 \times 3 \times 5 \times 5$

LESSON 8 (pp. 16 - 17)

1. a. 32 **b.** $32 \times 2 = 64$ **c.** $64 \times 2 = 128$
3. 243; multiplied 81×3
5. 1,280; multiplied 320×4
7. 10; divided $30 \div 3$
9. Answers may vary. Yianna doubled the previous term number. Oksana multiplied the previous 2 term numbers together.
Yianna: 2, 4, 8, 16, 32, 64
Oksana: 2; 4; 8; 32; 256; 8,192

LESSON 9 (pp. 18 - 19)

1. a. 34 and 55 **b.** 89
3. 36, 49, 64; Add the next consecutive odd number.
5. 55, 46, 48; Add 2, then subtract nine
7. 0, 2, 5, 12, 20, 30 **9.** 1, 6, 15, 20, 15, 6, 1

LESSON 10 (pp. 20 - 21)

1. a. Total Number of Cans: 1, 4, 10, 20, 35, 56, 84, 120, 165, 220 **b.** 220 cans
3. 110 **5.** 100

LESSON 11 (pp. 22 - 23)

1. a. 11 P.M. - 2 A.M. - 5 A.M. - 8 A.M.
b. 4 A.M. - 8 A.M. **c.** 8 A.M.
3. a. 1,800 **b.** 3,600 **c.** 5,400 **d.** $600 \times n$
5.

Hour	1	2	3	4	5	6	7	8
Times	10	20	30	40	50	60	70	80

Every hour Katrina does the exercise 10 times. In 8 hours: $8 \times 10 = 80$

LESSON 12 (pp. 24 - 25)

1. a. 1 hour **b.** 5 hours **c.** 10 hours
d. 20 hours **e.** 240 hours
3. Six years
5. 43,800 hours; students may use a variety of approaches.

LESSON 13 (pp. 26 - 29)

1. a. no **b.** yes

3. a. 5 **b.** 13

5. Yes. There is exactly one range value for each domain value.

For Exercises 7 - 9, tables will vary depending on the values chosen for x. Examples for values for x are given.

7. 1, 9; 2, 11; 3, 13; 4, 15

9. 0, 5; 1, 3; 2, 1; 3, –1

LESSON 14 (pp. 30 - 31)

1. Check students' graphs.

3. Check students' graphs.

5. –7, –5, –3, –1, 1, 3

7. Check students' graphs.

LESSON 15 (pp. 32 - 33)

1. a. 3

b. No. There are a number of shoe sizes that have more than one height. It fails the vertical line test.

3. not a function **5.** not a function **7.** function

LESSON 16 (pp. 34 - 35)

1. a. 0 **b.** 2 **c.** 4 **d.** 6 **e.** 8 **f.** 10

3. The graph slants up from left to right and intersects the axes at $x = -3$ and at $y = 3$.

5. a. They are parallel. They slant the same way.

b. The graphs intersect the axes at different points.

c. The number added to or subtracted from the variable x.

LESSON 17 (pp. 36 - 37)

1. (2, 2), (4, 6), (7, 3)

3.

Rotation	A	B	C
0°	(0, 5)	(3, –4)	(–2, 1)
90°	(5, 0)	(–4, –3)	(–1, 2)
180°	(0, –5)	(–3, 4)	(2, 1)
270°	(–5, 0)	(4, 3)	(1, –2)

LESSON 18 (pp. 38 - 39)

1. a. (5, 5) **b.** 6 spaces to the right

3. Triangle ABC: $A(7,2)$, $B(10, 6)$, $C(11, 3)$

5. Slide 2: $A(-5, 0)$, $B(-2, 4)$, $C(-1, 1)$

LESSON 19 (pp. 40 - 41)

1. a. A (2,6); B (5,3); C (1,1)

b. A' (6,–2); B' (3,–5); C' (1,–1)

3. a. Slide down 5 units, rotate counterclockwise 90°

b. Rotate counterclockwise 90°, slide right 5 units

5. Check students' work. **7.** $(-y, x)$

LESSON 20 (pp. 42 - 43)

1. a. 135° **b.** 180° **c.** 45°

3. a. 90° **b.** right

5. a. 2, 3, 540° **b.** 3, 4, 720° **c.** 4, 5, 900°

d. $n - 3$, $n - 2$, $(n - 2) \times 180°$

LESSON 21 (pp. 44 - 45)

1. a. $A = \pi \times r \times r$, or πr^2

b. $A = 3.14 \times 2 \times 2$ **c.** 12.56 cm^2

d. 6 cm; 113.04 cm^2

e. nine times greater

3. The area increased by the given factor.

LESSON 22 (pp. 46 - 47)

1. a. 25, 36, 64 **b.** 61 **c.** No

3. Yes **5.** Yes **7.** Yes **9.** 34

11. a. If you take the 3-4-5 triple and increase each number by the same factor, you get another triple.

b. Answers may vary. Possible triples: 12, 16, 20; 15, 20, 25; 18, 24, 30

c. Yes. Students should use $a^2 + b^2 = c^2$ to confirm their triples.

LESSON 23 (pp. 48 - 49)

1. a. $3 + 19 = 22$ **b.** $5 + 17 = 22$
c. $11 + 11 = 22$

For Exercises 3–20, two pairs are given. Answers may vary.

3. 5 + 19; 7 + 17 **5.** 5 + 23; 11 + 17
7. 3 + 29; 13 + 19 **9.** 5 + 31; 7 + 29
11. 3 + 37; 11 + 29 **13.** 3 + 41; 7 + 37
15. 7 + 61; 31 + 37 **17.** 5 + 79; 11 + 73
19. 3 + 97; 11 + 89 **21. a.** 4
b. In order to get an even number, you need to add another even number to the number.

LESSON 24 (pp. 50 - 51)

1. a. 3 + 3 + 19 **b.** 3 + 5 + 17
c. 5 + 7 + 13 **d.** 7 + 7 + 11

For Exercises 3–12, two possible answers are given. Answers may vary.

3. 3 + 7 + 19; 3 + 3 + 23
5. 3 + 7 + 23; 5 + 5 + 23
7. 3 + 3 + 31; 3 + 5 + 29
9. 3 + 7 + 31; 5 + 7 + 29
11. 3 + 5 + 53; 3 + 11 + 47

For Exercises 13-18, only one example is given.

13. 3 + 5 + 37 **15.** 3 + 3 + 59
17. 3 + 7 + 89
19. Answers may vary. Jody's conjecture seems to be true for all the examples in this lesson. The prime number 3 can be used with any of these numbers.

CUMULATIVE REVIEW ANNOS (L1–L4) (p. 52)

1. 33, 40, 47 **3.** 39, 48, 57
5. 6, 12, 18, 24, 30
7. 28, 35, 42; 210
9.

n	1	2	3	4
$5n + 4$	9	14	19	24

11. $4n + 2$; 362
13. $7n - 2$; 628

CUMULATIVE REVIEW (L5–L8) (p. 53)

1. 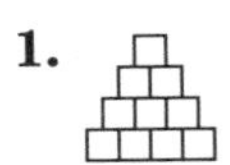

Add one more each time to each row than was added previously.

3. The numbers 13 and 23 should be circled.
5. $2 \times 2 \times 3 \times 7 = 84$ **7.** 50; half of 100

CUMULATIVE REVIEW (L9–L12) (p. 54)

1. 18, 24, 31, ...; Add one more than what was added before.
3. 6, 9, 14, 21, 30, 41, ... **5.** 62
7. 5 P.M.; 11 A.M.
9. The same day as today; 91 is a multiple of 7.

CUMULATIVE REVIEW (L13–L16) (p. 55)

1. Yes **3.** – 8, –4, 0, 4, 8
5. Check students' graphs. **7.** not a function

CUMULATIVE REVIEW (L17–L21) (p. 56)

1. Check students' grids
A (–3, 4), *B* (–2, 2), *C* (–5, 1),
D (4, 3), *E* (2, 2), *F* (1, 5)
3. Rotate 90° counterclockwise and slide up 5 units or slide right 5 units and rotate counterclockwise 90°.
5. 360° **7.** four times

CUMULATIVE REVIEW, (L22–L24) (p. 57)

1. b **3.** Yes **5.** 20 **7.** 51 **9.** b

For Exercises 10–13, there is more than one possible answer. Two are given.

11. 3 + 61; 5 + 59
13. 3 + 11 + 43; 5 + 11 + 41